Design Science and Innovation

Series editor

Amaresh Chakrabarti, Centre for Product Design & Manufacturing,
Indian Institute of Science Bangalore, Bangalore, India

The book series is intended to provide a platform for disseminating knowledge in all areas of design science and innovation, and is intended for all stakeholders in design and innovation, e.g. educators, researchers, practitioners, policy makers and students of design and innovation. With leading international experts as members of its editorial board, the series aims to disseminate knowledge that combines academic rigour and practical relevance in this area of crucial importance to the society.

More information about this series at http://www.springer.com/series/15399

Sushil Chandra

Aesthetics: Quantification and Deconstruction

A Case Study in Motorcycles

Springer

Sushil Chandra
Department of Body Engineering
Hero Motocorp Limited
New Delhi
India

ISSN 2509-5986 ISSN 2509-5994 (electronic)
Design Science and Innovation
ISBN 978-981-10-6234-6 ISBN 978-981-10-6235-3 (eBook)
https://doi.org/10.1007/978-981-10-6235-3

Library of Congress Control Number: 2017952004

Printed on acid-free paper

This Springer imprint is published by Springer Nature
The registered company is Springer Nature Singapore Pte Ltd.
The registered company address is: 152 Beach Road, #21-01/04 Gateway East, Singapore 189721, Singapore

I dedicate this book to my
granddaughter Sehr-e-nau (my new dawn).

Acknowledgements

The germ of this idea of quantifying aesthetics was put in my head by my Ph.D. supervisor. One day, during a casual conversation, he asked me, can we have a computer which scans an object and produces an output which describes the aesthetics of the object. This question set me to the path of two further questions, (1) is there an objective method to describe aesthetics? and (2) Can it be measured or calculated? Looking for an answer to these two questions led to this book. So, Prof. S.K. Atreya, Head of Faculty for Industrial Design at Indian Institute of Technology, Delhi is actually the progenitor of this book and no acknowledgement can start without mentioning him and thanking him.

Going through this book, you will realize that that the backbone of the research is provided by the 3D models and images generated by free-hand sketches. This would not have been possible, if my two friends Mr. Rahul Jain and Mr. Umesh Sharma would not have extended their helping hands and this page gives me an opportunity to express my gratitude. Moreover, this research was not possible without cooperation from the respondents. Though I would love to thank them by name, but the sheer numbers make it impractical. But this does not take away the fact that their contribution demands an acknowledgement of their contribution.

My own horizon on motorcycles has widened after I came in contact with The International Journal of Motorcycle Studies and this widening of my horizons has positively contributed to the scope of this book. I express my gratitude to the editors of this journal Prof. Suzanne Ferris and Prof. Steven Alford from Florida. Continuing with the mention of this journal, I thank the current editor Ms Sheila Malone who, without batting an eyelid, proverbially speaking, allowed me to use the materialfrom my articles published in the journal. Similarly I am thankful to my *alma mater* BITS, Pilani for allowing me to use the data from my Ph.D thesis.

But, most importantly I would like to remind my readers that I am not an academician and writing a book is unusual for a professional engineer. But this happened due to constant goading by two people in my life who did not allow me to rest till this happened—my wife Poonam and my daughter Nimisha. Though this book is an unusual place for thanking them but I cannot resist the temptation.

Contents

About the Author

Sushil Chandra is an R&D engineer working with Hero Motocorp, the world's largest motorcycle manufacturer. He was part of the design team for "Splendor", the world's best-selling motorcycle, and was chiefly responsible for the design of Hero Motocorp products, which included both studio and engineering aspects. He currently heads the body engineering function group, which vertically integrates all activities from conceptualization to mass production and focuses on the interplay of social, political, and philosophical aspects with the world of engineering design. Literature, especially ancient and modern poetry, happens to be one of his loves. This confluence of engineering and art drew him to the field of automobile aesthetics, which he found intriguing and exciting at the same time, leading him to complete a Ph.D. in the engineering of aesthetics. He has published several papers on the semantic aspects of motorcycle design.

Chapter 1
Introduction: Motorcycle and the Semantics of Design

Moonlight Sonata is a composition by Beethoven which, many believe, makes the blind experience the bliss of a moonlit night. Listening to this composition, it is difficult to believe that the experience can be put down in numbers. But we all know that it is very much possible to script the sonata in notations and notations are nothing but frequency and amplitude in numbers. In fact, it was Pythagoras (who firmly believed that everything in the world is numbers) who actually came out with the basic theory to translate music into numbers. But here we can challenge Pythagoras with the question whether we can put down in numbers the feelings generated by Moonlight Sonata. The obvious answers will be in negative but we all know that obvious answers are not always the right answers.

Returning to the question, we are living in the age of instant data capturing and analysis. You can record sound and your computer can immediately analyze the frequency and amplitude. You can scan any visual data and analyze the color, the geometry, the surface roughness, and any other data. Someone asked me, is it possible to capture and analyze the aesthetics through computer. The answer obviously is no but again the obvious answer is not necessarily the right answer and this set me thinking. If indeed we want computers to scan and analyze aesthetics, we must answer what are the prerequisites'? And more importantly, if at all this happens, how valid will it be?

First of all can aesthetics be quantified? Leave aside quantifying, can it be objectively described? By objective description, I mean, if a thousand people look at an object, will their description of aesthetics be same? Again the obvious answer is no? But let us assume the answer to both the questions is "yes". In that case, we need a grammar which defines the units and scale of aesthetics. But the real problem starts after that. Even if we have the parameters, units, and scale of aesthetics, we need a method to measure it. This needs an algorithm to convert the captured visual data into the parameters of aesthetics. Now, the captured visual data can only give you engineering parameters and the algorithm needs to translate them into aesthetic parameters. This book is all about developing that algorithm.

S. Chandra, *Aesthetics: Quantification and Deconstruction*,
Design Science and Innovation, https://doi.org/10.1007/978-981-10-6235-3_1

Now we are caught in a contradiction here. We all know that aesthetics is indeed subjective. Something which looks outright ugly to someone can be terribly exciting to someone else. And here we are talking about "objective description"! How do we reconcile the contradiction? Now, two things can clarify the doubt a bit. You can call it the philosophy of science. First, what we know as scientific objective reality is not really an objective but a subjective reality. For someone sitting in a train compartment another traveler sleeping on the facing birth is having a velocity almost nearing zero. But for a person outside the train, the same sleeping traveler is moving at a high velocity, equal to the train velocity. So the velocity of the person becomes a subjective reality. Still, the velocity of the traveler is a scientific fact. This means, everything that we consider an objective reality is actually a subjective reality depending on who is looking at it and the frame of reference. Whether it is the color of an object or weight, everything depends on the person, the construction of his eye, the light source, and the planet on which the object and the person exists. Now let us consider an election for the parliament of a country where different people vote for different parties. The party getting maximum votes becomes the most popular party and this is an objective reality irrespective of people's personal choices. These two arguments prove that objective and subjective realities are actually two faces of the same coin. What appears as objective is actually subjective and what appears to be subjective can be converted into objective reality. The point behind all this is that aesthetics may be subjective individually but it can be translated into objective reality by statistical means.

But our problems do not end here. The vocabulary of aesthetics is an open field. Poets to lovers to philosophers have described beauty of various objects in innumerable words. But, at the end of the day, all artistes, be they poets, movie-makers, fiction-writers, architects, fashion designers, or automobile designers are targeting to create emotions. So emotions are the building blocks or the vocabulary of aesthetics—across the fields of creativity. Once we are at the vocabulary, finding the alphabets is easy. Thankfully, the "Natya-Shastra" (The Handbook of Drama) by Bharat, which details all emotions and the circumplex of emotions by Watson and Tellegen combine together to provide the complete framework for finding these alphabets of emotion. The chapter on science of emotions details the science and mathematics for this quantification.

But, why am I talking all this in this book on motorcycles? After all I am an automobile engineer who has designed bikes throughout his life. And what do motorbikes have in common with music or art or philosophy? Now, motorcycle has always appeared mysterious to me. On one hand, it encompasses everything engineering and mathematics stands for and on the other hand it is an important aspect of art, culture, history, and politics. (Don't believe this? Please read Che Guevara's Motorcycle Diaries where his detour on Norton 500, we now know, changed the face of Latin America. Or ask Prof Suzanne Ferris and Steven Alford, professors in English Literature in Florida, who started the International Journal of Motorcycle Studies to profile motorcycling as a part of literature studies).

But motorcycles, as I said, are mysterious objects. And mystery lies in the inter-molecular spaces between engineering and art interwoven in it to create emotions. The visuals of a motorcycle are an initial temptation to the rider and the riding experience must fulfill the promise. The emotions generated by the visual must be realized once the rider actually rides the bike. The problem here lies in the fact that the visual architecture of a motorcycle is very complex. With all its functional parts visible, the multiplicity of parts and their range of characters make the visual design far more complex, particularly in comparison to cars. So it becomes a challenge for the designer to ensure that the performance aspects of a motorcycle correlate well with the visual design without dissonance between the two so that a coherent narrative is created.

1.1 Semantics of Design

The meanings and contours of design have been subject of fierce debate. Votaries of "form follows function" and "form follows emotion" have been arguing for quite some time. One positive outcome of this debate has been that the horizons of design in general and motorcycle design in particular have expanded beyond recognition. The traditional approach to design takes into account the functionality, the durability, the aesthetics, the manufacturing process and the cost into account whereas the modern approach goes into history, geography, psychology, sociology, and politics of the artifact besides the engineering. For example, if an engineer has to design a simple thing like glass, the designer will go into how drinking a liquid evolved over time, how it varies across the countries based on their cultural sensibilities and geographical needs, how the user interacts with glass socially and emotionally and how the glass features in the interplay between the different social, religious, and economic groups in a society. Chitturi et al. [1] define design as a trade-off between hedonic and utilitarian benefits and preventive and promotional emotions. In context of a motorcycle, this applies in an even more complexity. Today, the idea of critical design has emerged, informed directly by critical social theory in the sense that it is important to engage in the ways in which the world may change rather than purely observing the world as it is [2]. In a customer based business paradigm, an understanding of customer and market trends is driving the role of design that is involved in emotional dimensions of products and services [3]. Design has also been seen as a communication process where the artifact is the emitter and consumer is the receiver of the communication. But Crilly et al. [4] have criticized this notion and described design as triangular communication process between artifact, designer, and consumer. Schifferstein et al. [5] have described the consumer product attachment deriving from the product meanings relating to enjoyment, autonomy, group affiliation, and life vision. The semantic approach takes into account the meaning of the artifact as seen by the designer and the user. The meaning attached by the user is a product of three kinds of emotions—visceral, behavioral, and reflective [6]. Since this meaning as seen by the user is also to be understood by designer, it actually takes second order understanding (understanding of someone else's understanding) to complete the process of design. Though semantics as an approach to design got recognition by the end of twentieth century only, the evolution of motorcycle design has been largely semantic since inception. As Krippendorf [7] defines semantics as the science of meaning, he also categorizes meaning in three—meaning in use, meaning in language and meaning in ecology.

While considering the meaning of motorcycles in use as seen by the user, it amounts to (1) the interface of rider with the artifact, (2) the visual metaphors, and (3) the affordances of the artifact as a whole and in parts. More important in this context is second-order understanding of the artifact, as a whole, in parts and as parts constituting the whole. Besides the meanings in use that an object offers to the users, there is also a sub-text of communication, which decides the meaning of the object in language. Following aspects constitute the essential features of meaning in language:

Characters: Each artifact assumes a character in language. This character evolves through the social perceptions and collective memories of a society through its history. Companies strive to develop products that carry distinctive references to the character of their brands. Characters have following characters:

(a) They are different from properties in that they are creations of human perceptions.
(b) They are relative and comparable.
(c) They exist in opposites and continuums like hard-soft, heavy-light, robust-fragile, etc.

Category: This means how human beings categorize the artifacts as sub-category, category, and super-category. So far as motorcycles are concerned, it started as part of super-category of transport. When it started, automobile itself was a category which had to compete with other modes of transport like bicycles, horse carriages, etc. Within this, two wheelers became a sub-category. As times changed, automobiles became the de facto mode of transport and two wheelers became a category within the super-category of automobiles and motorcycles became a sub-category. Further down the line, motorcycles also had the sub-categories of cruiser, sports, and standard.

Identities: Users have different identities-personal, institutional, and social. In many cases, it depends on individual which identity he wants to bear and which he wants to be subsumed. The identity he wants to express through the artifact becomes the intended identity of the artifact. Today, motorcycle styling is almost completely synonymous with identity of the rider.

Narratives: Goldie [8] defines narrative as kind of representation of a sequence of events revealing certain coherence in what happened, configuring casual, and other connections into a narratable episode or episodes. Humans live in stories and the objective of designing an artifact is that a coherent story with interfaces of humans with artifacts is feasible.

Now, philosophers of design, over the years have fiercely debated the elements of design for a desired narrative. This debate has, so far, crystellized in two camps-the Gestalt theory camp and structuralist camp. Gestalt theory is based on Gestalt psychology, which believes that in visual human perception the whole is bigger than the sum of its parts and hence the overall perception of design cannot be deconstructed into various elements. Pratt [9], says the way the eye and mind actively manipulate what we see causes something to be aesthetically pleasing. As early as 1941, Campbell [10] argued that the effect of the whole over the sum of the parts is considerable as against the contention of Hollingworth and Foley who felt it is negligible.

On the other hand, structuralist thinkers such as Kotnik [11] define structure as a pattern resulting from a network of formal relationships between the elements of a set and a topological structure about the geometric organization of elements. To put it simply, the difference between the whole and the sum of the parts is also a mathematical construct and can be expressed in terms of mathematical relationships. But the dilemma of the structuralist approach lies in its inability to understand the process of transformation [12]. There have been attempts to resolve this dilemma with the help of ornamentality through rhythm. These attempts, though all these debates took place in the realm of architecture and urban planning, can be particularly seen in the design architecture of motorcycles, where a conscious attempt is made to create rhythm through the monotony of large number of components. The reality or the ultimate solution, as always, lies in the synthesis of these two extreme positions. Gestalts themselves can be treated as mathematical constructs and algorithms can be generated to understand the process of transformation.

1.2 Motorcycle and Aesthetics

Now if we look at motorcycles from this perspective, they started as a mode of transport without much concern for emotions. But one big problem with art is that it affects all facets of life, including engineering. This, coupled with the fact that the historical and technological developments made the motorcycles transform beyond recognition, brings in question the eternal debate of design between votaries of "form follows function" and "form follows emotion" in the realm of motorcycle design. This interplay of technology, history, and art has been detailed in the chapter on history of motorcycles.

Now, as long as we are taking a macro-view of aesthetics and motorcycles, it is easy sailing. Once we get into the actual deconstruction of motorcycle aesthetics, we have to enter the microterritory and here, life becomes difficult. We find a huge number of motorcycle parts, an infinite number of possible solutions for each part, a huge number of design factors responsible for emotions and an infinite number of possibilities for each factor. If we combine them together we are faced with an impossible situation. Another difficulty is prioritization. All design factors and motorcycle parts are not equally important. In this situation, it becomes important to devise a methodology that accounts for their weightage matching with their relative importance. This is an area dealing with the nuts and bolts of the process of navigating through the ocean and has been dealt in the chapters on design factors and methodology. End of the day, we end up with a solution where we can quantify each emotion for each motorcycle part for each design factor and the method to synthesize all of them.

This was the operational part. But, once the emotions appear in form of numbers, they start telling their own stories and these stories have layers and layers like the labyrinthine corridors of a conspiratorial castle. The most prominent revelation out of these stories is that the design factors themselves do not contribute that much to the emotions as the interplay between the factors like the interplay between color and texture, the interplay between typography and graphics and the interplay between unity and dynamism. And the most dominant determinant of the emotion matrix is the interplay between unity and dynamism. But, this is only the tip of the iceberg and in fact each factor has its own story and each story has been told in respective chapters.

Now, the test of any pudding lies in eating and the test of any algorithm lies in its testing. Having developed the algorithm it needs to be put to test. Relatively, it is easy. We have to collect the actual emotions directly from people and compare it to the calculated results. Not surprisingly, there are no surprises, proving the efficacy of the algorithm. But since this book is more about methodology and less about results, more important point is the process of closing the loop by verification. Not necessarily, the results will match in all cases. In case they do not match, the necessary tweaking of results should be done.

Finally, what we are dealing here with is aesthetics and emotions which invariably are linked with culture and culture, as we know, is dynamic and has many aspects. Point is not to explain that emotion scores have to be re-established in every cultural context but to discuss the mechanisms and levers through which culture affects aesthetics. The last chapter discusses these levers and mechanism of culture.

Finally, let me declare a disclaimer. Though this book has established quantitative emotion score for design drivers, it should be very clear that these score have been determined only in a specific cultural context. Fundamentally, this book establishes a framework for quantification of aesthetics and derives a methodology based on the framework.

References

1. Chitturi, Ravindra. 2009. Emotions by design: A consumer perspective. *International Journal of Design* 3 (2): 7–17.
2. Rust, C. 2007. Unstated contributions-How artistic inquiry can inform interdisciplinary research. *International Journal of design* 1 (3): 69–76.
3. Kim, Bo-Young, and Bum-Kyu Kang. 2008. Cross-functional cooperation with design teams in new product development. *International Journal of design* 2 (3): 43–54.
4. Crilly, Nathan, Maier, Anja and Clarkson P.John. 2008. Representing artifacts as media: Modeling the relationship between designer intent and consumer experience. *International Journal of Design* 2 (3): 15–27.
5. Schifferstein, Hendrik N.J, and Zwartkruis-Pelgrim, Elly P.H. 2008. Consumer-product attachment: Measurement and design implications. *International Journal of Design* 2 (3): 1–13.
6. Norman, Donald A. 2005. Emotional design, why we love (or Hate) everyday things. USA, Basic Books, 1: 21–24.
7. Krippendorff, Klaus. 2006. *The semantic turn, a new foundation for design*, 39–175. New York: Taylor and Francis.
8. Goldie, Peter. 2009. Narrative thinking, emotion and planning. *The Journal of Aesthetics and Art Criticism* 66 (2): 97–106.
9. Pratt, Richard, J. 2011. How perception works and its role in creating engaging design. In *Design Principles and Practices International Conference at Sapienza*, Italy: University of Rome.
10. Campbell. I. G. 1940. Factors which work towards unity or coherence in visual design. Elsevier, 146–162.
11. Kotnik, Toni. Algorithmic design: Structuralism reloaded. 327–335.
12. Cruickshank, Leon. 2011. The case for a re-evaluation of deconstruction and design; against Derrida, Eisenman and their choral works. *The Radical Designist* 1–9.

Chapter 2
Motorcycle and Its Aesthetics: A Glimpse in History

History is but a confused heap of facts
—Chesterfield

Therefore, looking at history, any history, is a complicated exercise. If it would have been just listing the chronology of events, it would have been easy. To be honest, even that is not easy. In fact, at any point of time, hundreds of incidents happen in many parts of the world. To judge the incidents, to select the ones worth a mention and to link them through a narrative is a daunting task. But this is not what we mean by looking at history. Because, when we do this, we have to decide the insight that needs to be derived. To decide the insight, we need to answer this question-why at all we need to look at the history of motorcycles in a book discussing motorcycle aesthetics. You can guess the answer is not very difficult. The parameters of aesthetics are inextricably linked to the social sensibilities across the world and the social sensibilities traverse diverse trajectories in different societies. The evolution of these sensibilities depends not only on the histories of societies themselves but the evolution of cultures and technologies as well. Moreover, the parameters of semantics themselves evolve with history. So, the insights that we want to derive from history are twofold:

a. The relationship between the technological evolution of motorcycles and social and cultural evolutions.
b. The evolution of semantic parameters linked to motorcycle technology.

So the contours of the discussion are getting clearer. Like all histories, we will divide the history of motorcycles in periods and each period will be discussed across different geographies. Each of these small pieces will be discussed for the technological evolution, the impact of social and cultural changes on them and the semantic parameters.

S. Chandra, *Aesthetics: Quantification and Deconstruction*,
Design Science and Innovation, https://doi.org/10.1007/978-981-10-6235-3_2

2.1 1869 to First World War: The Birth-Pangs

Europe was the center of global awakening in this period and America was on the cusp of greatness. The decline of Vatican in the affairs of state and feudalism shedding the space to capitalism had heralded new ideas in areas of science, arts, and literature. Capitalism was a fresh flower free from feudal prejudices and goading people to invest in new ideas. At the same time, industrial revolution, as the industry was moving from craft to mass manufacturing, for the first time created the necessity for people to commute long distances on a daily basis. Moreover, the frequent Napoleonic wars had caused the fodder prices to go through the roof and personal transport using horses and horse carriages became unaffordable. So, the three factors—industrial revolution and scientific awakening, the shift from craft to mass manufacturing and high fodder prices—made the situation ripe for invention of a cheap, fast and automated (not animal driven) mode of transport for lower income group.

The invention of motorcycle was, of course, the logical consequence of two great breakthroughs—the internal combustion engine and bicycle and obviously the starting visual metaphor was exactly the reflection of that (Fig. 2.1; Table 2.1). The visual metaphor of motorcycle was clearly derived from bicycle with an engine added to that.

Before we talk about the semantics in this period, we can have a look at the causative factors—cultural and socioeconomic influences leading to the semantics. This was the time when Europe was abuzz with Marxist thought in midst of

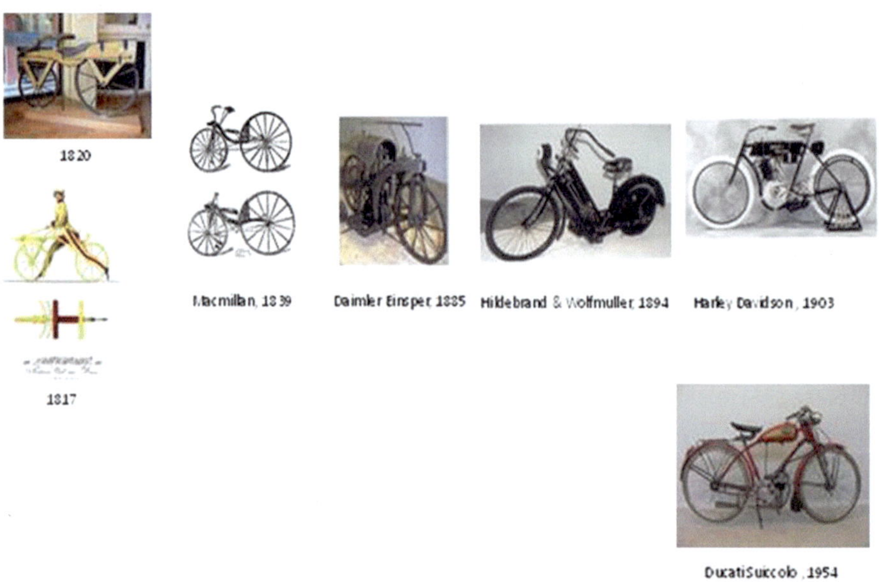

Fig. 2.1 The evolution of the visual metaphor

Table 2.1 Engineering specifications

Specification	Daimler Einsper (1885)	Hildebrand and Wolfmuller (1894)	Harley Davidson (1903)
Engine	Single-cylinder air cooled, inlet over exhaust two valves	Water cooled four-stroke parallel twin	Air cooled, inlet over exhaust two valves
Capacity	264 cc (58 × 100 mm)	1489 cc (90 × 117 mm)	412 cc (76 × 89 mm)
Max power	0.5 hp@600 rpm	2.5 hp@240 rpm	4 hp
Transmission	Single speed belt final drive	Drive to rear axle	Single speed shaft final drive
Frame	Wood	Tubular steel	Steel twin down-tube
Suspension	None	None	Telescopic front, rigid rear
Brakes	None in front, shoe in rear	Friction spoon in front, bar in rear	None in front, contracting band in rear
Top speed	7 mph	28 mph	40 mph

industrial revolution and "the spectre of communism was haunting Europe"—to quote the Communist Manifesto. This was not an isolated influence. From Architecture to literature to fine arts—every field of cultural life was seeing an awakening of progressive thought with protest against the old order and the downtrodden at the center of the cultural process. The two dominant streams both in literature and fine arts were romanticism and realism—both expressions of revolt against the elitism and classicism of earlier ages. Whereas the romanticist poetry of Wordsworth and Coleridge portrayed the life and times of shepherds and grass-cutters, the realist fiction of Mark Twain and O' Henry chronicled the lives of people at the margins of society.

The life and struggles of the common people (in both the romanticist and realist school of art and literature) juxtaposed against the classicist tastes of neo-classicist school, was common thematic thread of this period (Figs. 2.2 and 2.3). The same evolution of consciousness was evidenced in architectural evolutions in the form of neo-renaissance school (Fig. 2.5) and gothic revival (Fig. 2.4). Both these schools were predominantly the product of the new consciousness. At the level of structural themes, both these schools of architecture emphasized on simplicity of shapes in form of triangles, rectangles, and arcs. At the level of literature as well, the emphasis was on simple narrative structure in fiction and simple rhythmic style in poetry. As we will see through the design language of motorcycles of this period, these thematic and structural influences were present despite the technological limitations in the field of manufacturing process as well as the fundamental physics of engines and dynamics.

Gottlieb Daimler joined Gasmotoren Fabrik Deutz, a company owned by Nicolaus Otto making stationary engines based on Lenoir cycle, as factory manager in 1872. In 1876, Otto and Daimler together created the compressed charged cycle,

Fig. 2.2 The realist art of nineteenth century. Third Class Carriage by Domier (*left*) and painting by Millet (*right*)

Fig. 2.3 Romanticism in art. "Liberty" by Delcroix

known as Otto Cycle. After this invention, two more important inventions took place. Otto created the Magneto to create the spark and Daimler invented the Carburetor with Wilhelm Maybach to mix fuel with air. In 1883, Otto and Daimler developed differences. Otto wanted to continue making stationary engines and Daimler wanted to apply the concept for transportation with small engine. The differences got to a level where they parted ways and Daimler took Maybach with him to establish another company, which is today known as Daimler-Benz. This pair of Daimler and Maybach, while making the initial concept for the car, added two wheels to the engine and created the first prototype of a motorcycle. (see Fig. 2.1, Daimler Einsper).

The first production motorcycle, however, was designed by Hildebrand and Wolfmuller and launched in 1894. The inspiration had been taken from a steam

Fig. 2.4 Gothic revivalist architecture. *Note* the linear structure (where length is considerably more than width) and *triangular* shapes formed by *double* arches

Fig. 2.5 Neo-renaissance architecture. *Note* the *rectangular* structures (where length is almost equal to width) and *semi-circular* arches

powered design by Hildebrand brothers, where the rear wheel was directly driven by connecting rod and the rear mudguard doubled as water tank for engine cooling (see Fig. 2.6).

This was the time, when three iconic brands were taking their first steps—Harley Davidson, Indian and Royal Enfield. After making many attempts, Royal Enfield made its first V-Twin engine bike in 1910 in Birmingham. Harley and Davidson while attempting to make a motored boat created their first motorcycle in Milwaukee (Fig. 2.1). And George M. Hendee, who was making bicycles in his company from 1897, teamed up with a fellow bicycle racer Oscar Hedstrom to produce a 1.75 bhp single-cylinder bike in 1900 and by 1916, they made their V-twin motorcycle Powerplus (see Fig. 2.7).

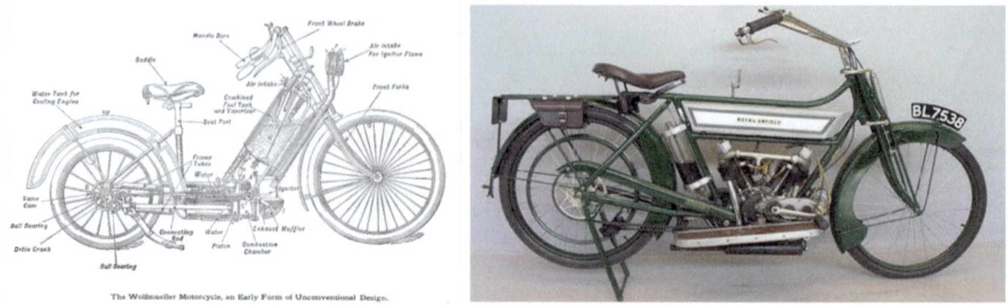

Fig. 2.6 Hildebrand and Wolfmuller

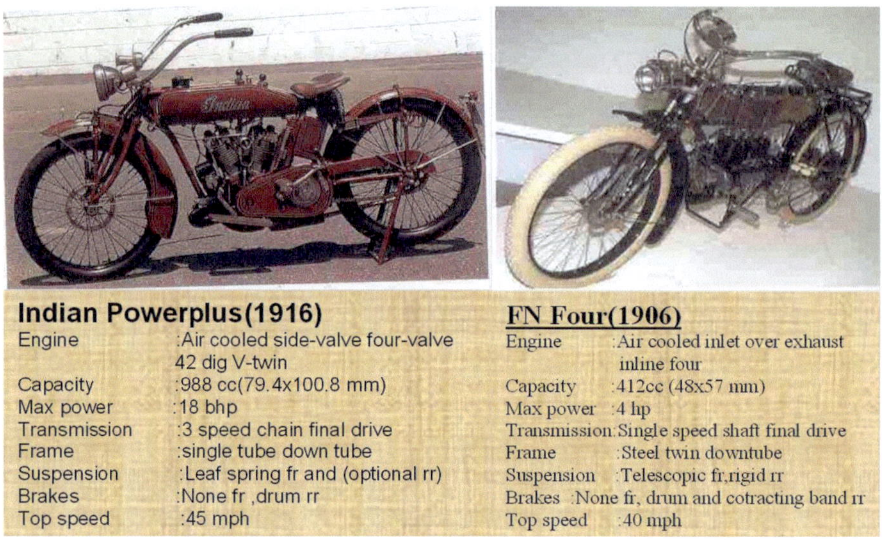

Fig. 2.7 Indian and FN Four

Considering the fact that four-stroke Otto cycle was invented only in last quarter of nineteenth century, it is amazing that by the end of first decade of twentieth century four-cylinder engines were being made for motorcycles and the maximum speed raced from 7 to 45 mph in such a short time and both water and air cooling was available for cooling purpose. For transmission, it was either belt drive or chain drive as geared transmission had not arrived at that time and speed variation was achieved by multi-sprocket chains. Most noticeable by their absence was modern braking and suspensions, though lights were added at the fag end of the first decade. Seats and wheels were direct imports from bicycles. Another important feature was extensive use of sheet-metal forming for body work, considering the fact that sheet rolling technology was introduced only around 1850 and forming was introduced only toward the end of nineteenth century. Extensive use of tubes was natural keeping in view the ancestry from bicycles.

2.2 Between the Wars: The Baby Steps

The period between the two wars is known for two critical events. The first is the occurrence of the two wars themselves and the second was peaking of social unrest in Europe resulting in two exactly opposite streams of social tendencies. On one hand, the quest for economic equality reflected in Soviet Revolution leading to spread of socialist thought throughout the world, on the other hand, it gave rise to right wing fascist thought which ultimately resulted in the Second World War. But technologically the most important change was the settling down and scaling up of mass production. This effect was augmented by the technological advances due to the wars like hardening of metals.

The field of art and literature saw the onset of three important schools—expressionism, cubism, and surrealism (Fig. 2.8). Expressionism is a school where an artist or a writer seeks to express the inner world of emotions rather than the external realities. We will see in the later chapters of the book that this school of art had the most profound impact on automobile design as emotions became the most important aspect of design language. Cubism, the most prominent exponent being Pablo Picasso, was to become a very prominent harbinger of things to come with its emphasis on simple geometric shapes, as the modernist movement in design was actually the continuation of this thought. The first motorcycle made by BMW was actually an expression of cubism in design. Surrealism, Salvador Dali being the most celebrated proponent, was the school which was in fact realism with a surprising element. All these three art movements, along with the fourth one, impressionism (which laid emphasis on use of primary colors) had a profound impact on motorcycle design which saw the experimentation with colors with gay abandon.

The visual and engineering architecture of motorcycles saw a revolutionary change in this period- scooter branched out as a completely new category of two-wheelers. The most important change was in the visual metaphor. As the bicycle had derived its visual metaphor from horse and motorcycle from the bicycle, the bike retained the metaphor of a horse. But the scooter will have none of

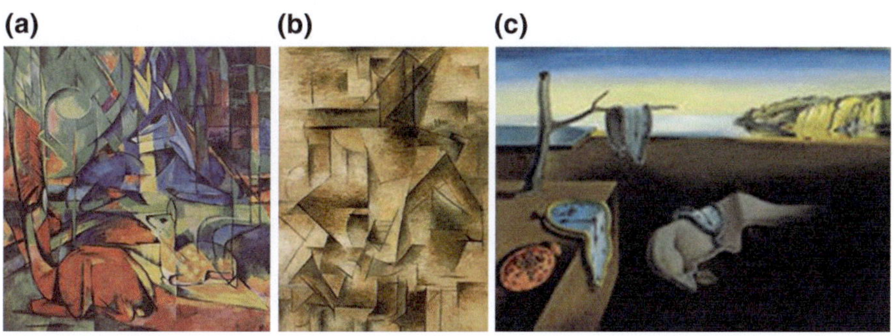

Fig. 2.8 **a** Expressionism by Franz **b** Cubism by Picasso **c** Surrealism by Dali

that and derived its visual metaphor from a chair where the user is sitting comfortably with his legs resting in front.

The most important brand that came out during this period and is still holding strong was BMW. It was a company by the name Rapp Motorenwerke, making aircraft engines and had to stop making them due to the restrictions put on Germany after the treaty of Versailles in 1919. As a result the company started making, besides other things, engine for Helios motorcycles (Fig. 2.9a) by a company Bayerische Flugzeugwerke. Later on, BMW acquired this company and finally in 1923, it introduced its first product called R32 (Fig. 2.9b) with an across the frame boxer engine designed by Max Friz. The engine was unique in the way that it featured a recirculating wet sump oiling system with a drip feed to roller bearings. Not only this, it was unique in its frame architecture and visual architecture as well with its double down-tube frame forming its visual style as well. The style statement was well ahead of its time with cubist design as the main attraction. So far as performance is concerned, it was the first bike to develop its maximum torque at crankshaft speed as low as 3200 rpm and achieve a maximum speed of 55 miles per hour.

This was only the beginning of the post-war technological explosion in motorcycles. Many new names appeared on the scene with hitherto unseen power and speed. George Brough from Nottingham came in with Brough Superior SS100 at 100 mph in 1925 fitted with drum brakes in front and rear, girder fork suspension in front, double spring suspension in rear and a four speed transmission. Another brand Triumph from England appeared in 1938. But, the pioneer Harley, by this time established its DNA of cruiser with a trademark V-twin engine with its typical exhaust sound. But the most remarkable aspect of this DNA was the design language with minimalism and slow smooth curvatures at its core (see Fig. 2.10).

The highlight of this era was the technological evolution as result of first war on one hand and the increasing interplay between art, emotion, and design on the other hand. So far as the ecological evolution is concerned, scooter branched out as separate category of personal transport and even among motorcycles, cruiser evolved as a category of motorcycles.

Fig. 2.9 **a** Helios motorcycle **b** BMW R32

Fig. 2.10 Brough, Harley, and Triumph

2.3 The Metamorphosis (1945–60)

The period after the war saw the scene completely turned on its head. Till now the whole game was being played out in Europe and US. After the war, as if by the flip of a coin, these actors exited the stage and new actors, mostly from Japan entered in a big way. All this happened in a short span from 1945 to 1960, when BSA, the manufacturers of Triumph brand, surrendered their position of world's largest motorcycle manufacturer to Honda. What actually happened that changed the game so suddenly?

What actually happened was that this period brought a revolution in the way motorcycles were perceived due to different political and sociological factors. Whereas the end of the war saw US emerging as one of the poles in a bipolar world, Roosevelt's new deal put money in common people's pocket. This factor gave birth to baby boomers' generation, where people wanted best of life for their children. This period saw cheap cars replacing motorcycles as the main mode of personal transport and motorcycle became a means of thrill, adventure, and enjoyment for the baby boomers. This was the period when motorcycles became a social and emotional statement as a result of cult movies like "The Wild One" and "Easy Rider." This saw the emergence of giant cruiser and sports bikes dominating the scene.

Another related development was the economic revival of Japan in post-war effort by US. This resulted in the emergence of biking giants like Honda, Suzuki and Yamaha, who eventually dominated the late twentieth century. Whereas BMW continued experimentation with earl fork and Ducati made a humble beginning with Cucciolo, it was Moto Guzzi in Italy which made daring experiments with color. The stylistic influences in this period mainly came from modernist architecture (Fig. 2.11) where clean lines with high use of technology ruled the roost.

Fig. 2.11 Modernist architecture. *Note* the simple geometric shapes with modern materials

Soichiro Honda, an auto repair shop owner speed maniac in Hamamatsu built a race car himself to participate in car racing. In 1936 he got injured in one of the races, formed a company to produce piston rings and started supplying to Toyota. After the war, when Japan was rebuilding itself, Honda sensed a need for affordable transport and started grafting two-stroke engines on to bicycles. Encouraged by the success of these machines, he started Honda Motor Company in 1948 and in 1949, brought the first machine, the model D (Fig. 2.12 left) with a 100 cc two-stroke engine. It was a complete departure in visual architecture from its western cousins. This was the first proper step-through motorcycle where the fuel tank was placed under the seat. Then after making model E with a four-stroke 125 cc engine and Honda cub with a 50 cc engine, finally Honda returned to normal step-over architecture with a 250 cc parallel twin (Fig. 2.12 right) in 1960. The highlight of this journey was Honda's obsession and success with racing which afforded very high reputation internationally.

Yamaha Motor Company started as an offshoot of Yamaha Corporation in 1955 with YA-1 (Fig. 2.13 left), a copy of German bike DKW-1RT-125 with a single-cylinder air-cooled 125 cc engine and coil spring suspension in front and

Fig. 2.12 Honda Model D (*left*) and 230 cc Parallel twin (*right*)

Fig. 2.13 Yamaha and Suzuki starters

rear. There was nothing new in engineering or visual architecture but the bike excelled at national and international races. The company continued with YA series and later started YD and YDS series for 250 and 250 cc performance engines.

Suzuki, a textile company established in 1909, started experimenting with engines clipped on to bicycles and came out with Diamond Free (Fig. 2.13 center) with double-sprocket wheel mechanism for two-speed transmission in 1953 and in 1955 improved upon to make Colleda COX (Fig. 2.13 right), a bike with a down-tube frame and an overhead valve 125 cc four-stroke engine.

As we can very clearly see from these events that this period saw a major shifting in motorcycle manufacturing from west to Japan and this brought about the shift from heavy multi-cylinder high capacity bikes to small single-cylinder bikes with major changes in visual architecture and semantics.

2.4 1960–1984: Decline of the Empire

As the baby boomers generation matured in US, oil crisis hit the world and the interest in motorcycles waned in US and Europe. Concurrently Japan established itself as the new automotive hub of the world and became the leading player in bike manufacturing, with Ducati and Harley barely holding on to their positions with Ducati Scrambler and HD Electra Glide. But it was predominantly a show by the Japanese quartet. Yes, the fourth of the Quartet, Kawasaki Aircraft started manufacturing motorcycles after taking over Meguro Manufacturing in 1962 with a 250 cc OHV engine. But this period is known for consolidation of Honda. It firmed its grip on 125 and 250 cc segments by winning races continuously, expanding its manufacturing bases in west and sweeping the American market with its campaign —“you meet the nicest people on Honda”—for its step-through bikes. This was the time when bikers had become infamous in US with self-inflicted badges of honor like “one-percenters” and “outlaws” due to biker-gangs and gang-wars immortalized by the movie “The Wild One.” Eventually, Honda ventured into multi-cylinder arena with 6-cylinder 250 cc bike 3RC164 (Fig. 2.14 left) which was an engineering marvel fitted with disc brakes for the first time. And then, in 1968, it made a big splash, both in market and race tracks, with the first mass-market four-cylinder CB750 (Fig. 2.14 center), again with disc brakes. And again in 1974, it made its

Fig. 2.14 Honda: the developing into banyan tree

first foray in tourer segment, till now dominated by HD Electra Glide with Goldwing GL1000 (Fig. 2.14 right), having the first water cooled engine fitted with a fuel pump. This was a vehicle with completely new engine architecture, new vehicle engineering architecture and a conventional look. The trick was to put the fuel tank below the seat and use the conventional fuel tank position as an electronic bay.

But, why have I chosen this period of 1960–84 as a period worth a separate mention. And what is the intriguing significance of 1984? For a person looking at motorcycle history from India, 1984 is special year, the significance of which, we will discuss later in this chapter. But somewhere in the first half of this decade, motorcycle design changed gears. Important point to remember is that the seeds of this top gear were sown in this period when the ecological diversification of motorcycles had taken shape. The categorization in commuter, cruiser, tourer, and sports had taken place. Another significance of this period was the consolidation of Honda as the undisputed leader so far as numbers and spread are concerned.

2.5 Post 1984: Return of the East

As US became the undisputed leader of post-cold war unipolar world, it decisively shifted to service economy. As the growth rates of western economies remained almost stagnant, India and China became the fastest growing economies. As the Japanese giants started looking for spreading their wings in Asia as their own demands shrank, they found readymade hungry markets in India, China, Thailand, and Indonesia. As all of them are third world emerging economies, motorcycles were predominantly for commuting in these countries. This resulted in the bike world getting polarized between western market (including Japan) with higher capacity cruiser and sport bikes and Asian markets with lower capacity commuter bikes. The US motorcycle industry saw a revival of sorts with the return of the baby boomers and the rise of motorcycling as a subculture. The recession of 2007 was another factor which helped the two-wheeler to make a comeback in Europe.

Fine arts, literature and architecture (Fig. 2.15) are dominated by minimalist (best illustrated by Hemingway's hard and metallic style of writing), modernist (simplification of form without ornamentation), and post-modernist (reaction to modernism with return to glamor and ornamentation) schools. The portrayal of motorcycle in movies saw an interesting twist. Instead of signifying defiance and

Fig. 2.15 Minimalist (*left*) and Postmodernist (*right*) architecture

Fig. 2.16 Movies: matrix reloaded (*left*) and mission impossible (*right*)

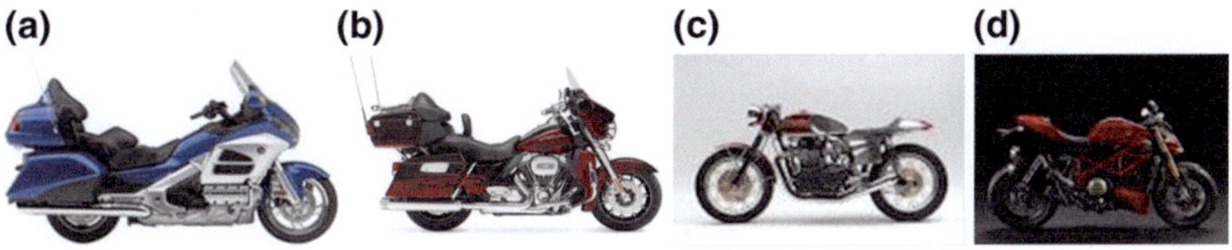

Fig. 2.17 **a** Touring Honda Goldwing **b** Touring Harley Electra Glide **c** Cafe-racer **d** Street-fighter

tryst with raw nature (the predominant trend in immediate post-war period), motorcycle identified more with speed and raw aggression (Fig. 2.16). As lyricism was lost somewhere in these movements (particularly in field of poetry), motorcycle styling saw an interesting phenomenon. It adopted the above schools with fresh whiff of lyricism, made possible due to advances in technology like CAD-CAM and plastics.

The two most dominant trends in this period are the proliferation of categories and explosion of stylistic schools. The high initial torque motorcycles expanded into cruiser, off-roader and tourer and high high-speed torque category virtually exploded into street-fighter, café-racer, and racing bikes (see Fig. 2.17). The commuter, which was earlier existing in step-over and step-through versions

Fig. 2.18 Minimalist styling (*left* and *center*) and Post-modernist styling (*right*)

developed into micro-segments. Technologically engine development saw a new high with highest ever bhp/cc of engine size.

The minimalist and post-modern schools of art and architecture have a profound influence on the styling of bikes (see Fig. 2.18). The minimalist school has found expression both in small size commuter bikes as well as high-powered naked styling of racing and cruisers bikes. One of the finest examples of minimalist styling was Harley's V-Rod, where Harley changed its visual architecture matching its brand-new engineering architecture but maintained its DNA of minimalist styling found. On the other hand, the post-modern sensibility is prevalent in full-cowl and unified body styling.

Talking about the brands, very clearly the global brands are divided into two categories—niche brands having a very narrow product line with strong company DNAs like Harley, Triumph, BMW, Ducati and Royal Enfield and Umbrella brands like Honda, Yamaha, and Suzuki having presence in all product lines. Another dominant trend of this period is the continental drift. With the center of gravity shifting to India almost all global brands like Harley (which prided itself for its true-blue Americanism), Indian, and Triumph scrambled to have a piece of the Indian pie. Royal Enfield is another category, which downed its shutters in UK and shifted to India with an Indian ownership. We will discuss more of the Indian story in the section, well, "The Indian Story".

Talking about the semantics of this period, the story is simple. With categories to cater to all possible emotions, there are as many narratives as there can be. And there are bikes to cater to each of the narrative.

2.6 The Indian Story

Though motorcycle manufacturing started in India only in 1961, with Indian Army placing an order to Royal Enfield on the condition that it sets up a manufacturing facility in India, the history of Indian popular culture has the seeds of future motorcycles in twenty-first century. As India leads motorcycle manufacturing quantitatively today, motorcycles are increasingly being designed with Indian sensibilities in mind. Till 1984, basically three brands Enfield, Rajdoot, and Yezdi

Fig. 2.19 Bullet (*left*) Yezdi (*center*) and Rajdoot (*right*)

(see Fig. 2.19) dominated the Indian scene. Henceforth with advent of lower capacity motorcycles, there was a dramatic shift as far styling goes.

From 1961 to 1984, same three vehicles ruled the roost, with miniscule quantities. But in the post-1984 period, Indian motorcycle matched the global styling quotient neck to neck. Why? In the period from 1960 to 1984, the motorcycle designing was shifting its center of gravity from US and Europe to Japan, whereas Indian motorcycles were completely depending on British and Czechoslovak designs. Moreover, since the quantities were miniscule, Indian sensibility was not catered to. But in post-1984 period, as the quantities increased dramatically and manufacturers (mostly aligned to Japanese global modernizers) kept their ears to the Indian ground. These factors coupled with the advent of CAD–CAM and new plastic materials, resulted in breathtaking array of lyrical designs.

There is another interesting sidelight to the Indian scenario, which is largely similar in South Asia and extends to South East Asia. While the global motorcycle designs have largely concentrated on high capacity bikes based on the concept of fun an adventure, motorcycle in Asian (when we say Asia, please exclude Japan for the current context) part of the globe, for all practical considerations, remained a medium of transport. Considerations of economy have restricted these motorcycles to capacities from 100 to 250 cc. Though, Asia has segmented the markets into power and economy bikes, even the power bike capacities are miniscule compared to global standards. Since, the hugest volumes are coming from Asia, the sensibilities have shifted from high street Western sensibility to main street Indian sensibility, which, as we have argued earlier, are essentially lyrical. With ready help coming from technology in form of CAD-CAM, flexibility of design and materials, the last barriers on road to lyricism are being lifted.

In the immediate aftermath of India getting freedom, Indian society was divided along three types of hierarchies-administrative, caste and feudal hierarchies running along identical lines, i.e., the people at top of the three hierarchies were the same. Similarly, People at the bottom of the three pyramids were also the same. In this period, cars were meant only for the select few among the urban rich. Motorcycles were meant for the rural rich and those forming the creamy layer of the urban middle class. This resulted in the fundamental social codes of rural dominance and urban quest for grace and class for motorcycle owners. In 1970 movie "Andaaz," the chartbuster song has boy driving a Bullet 350 with the girl in pillion seat in sidesaddle posture (Fig. 2.20 left). This inhibited expression of love finds

Fig. 2.20 Movies: "Bobby" (*left*) and "Sholay" (*right*)

expression in flat contours of seat and fuel tank and horizontal muffler. In the path breaking movie Bobby (1974), the girl, who claims to belong to twenty-first century, sits in parted legs position facing the front and nibbling the earlobes of the boy (Fig. 2.20 right). Now, the motorcycle was Rajdoot GTS 175. (interestingly supposed to be made especially for this movie, hence nicknamed "Bobby Rajdoot") with smaller wheels and long handlebars. Though the seat and tank contours remained flat due to technological constraint, muffler was made inclined to give a relatively dynamic look. This image was a precursor of future social behavior, when this sitting posture became common in metros.

The period from 1984 to 2000 was a watershed both for post-colonial India and the motorcycle landscape of the country. After her comeback from near oblivion, Indira Gandhi, having discarded her socialist pretensions, opened the windows for foreign technology, leading the Japanese giants to open shops in India. Hero-Honda, Escorts-Yamaha, TVS-Suzuki, and Kawasaki-Bajaj were the joint ventures as the result, which brought 100 cc fuel efficient bikes to India. At the same time, a much deeper development was taking place at sociopolitical level. The feudal, caste and administrative hierarchies, instead of remaining parallel as ever, started crossing each other as the aspirations of the people on the fringes started coming to the fore. With the arrival of the Maruti car, cars became affordable by upwardly mobile middle class. Inter-gender discourse became more forthright and upfront, at least in urban areas. The way a boy and a girl interacted with the motorcycle, was the extension of the Bobby phenomenon, with the difference that now, motorcycle was a willing participant in the process. All this social and moral upheaval produced another set of fundamental social codes.

The period after 2000 is significant on many counts. For such a populous country, we cannot expect gender equations to change significantly in such a short time. But the attitude of urban middle class certainly changed, at least in appearance. The social upheaval in the hinterland consolidated itself and the new power equations brought the marginal sections to the forefront resulting in increasing motorcycle penetration in country side. This was further helped by increased road coverage in countryside.

Today, having overcome the global recession, the Indian motorcycle manufacturing is on the cusp of the proverbial leap waiting for the direction that the economy takes. But whatever the direction of the economy some trends for the industry can be predicted with reasonable confidence:

a. The major chunk of the motorcycle population will always remain the commuter with the small size single-cylinder engine.
b. There will be shift to alternatives of fossil fuel keeping in view the limited global stocks and environmental pressures.
c. The lyrical sensibility of Indian civilizational memory will keep the styling of the bikes on the track of high unity design and high saturation colors with curvaceous surfaces.
d. As India becomes the hub of global styling the visual architecture and metaphor will see a major shift.

2.7 Summary

Looking at the whole journey of motorcycles some major trends can be spotted. With the West shifting to service economy, the hub of motorcycle manufacturing shifted to India and China. The collateral impact of this was a shift from modernist to post-modernist styling. At the same time, the global and niche brands have maintained a preference for both minimalist and post-modern.

The technological evolution in engine design has been in two directions—packing more power in every cubic-centimeters to achieve ever-improving performance and improving fuel-efficiency. One of the major reasons for this has been the increasing intervention of digital technology. The third dimension of reducing emissions has been added recently. The improvement in manufacturing technology has brought better fit and finish, durability and efficiency.

There has been a proliferation of categories with expansion in visual architecture and visual metaphor. Today the visual metaphor of bicycle coupled with an engine and a fuel tank has become extinct. Moreover, there are as many categories as there are narratives. Merging of the experiential and visual narrative has been the highlight of this story.

Coming to the fundamental question, what do you derive from this history to resolve the question of deconstructing the aesthetics? One thing has come out loud and clear. Emotions are an essential part of the civilizational sensibility. On one hand, the art, literature, and architecture of the times always impact the visual and experiential narrative. At the same time, the civilizational memory of the society also plays a vital role. The interplay of these components can be clearly studied only if we are able to deconstruct the emotions. The next chapters will take us there.

Chapter 3
Challenges for Product Planners

A product planner, take any organization, invariably thinks he is stuck in the most thankless place. Whether a product fails or succeeds, he is doomed in any case. If a product hits the bull's eye, everyone from R&D to manufacturing to sales and marketing rushes to steal the limelight. And they are right, because no product can succeed without all the three arms working for it. But if it fails, invariably, the failure is put at the product planners' door. The irony gets more complicated for the product planners as they have to deal with a world of ambiguities and uncertainties. Before we discuss these ambiguities and uncertainties, let us first understand who are these product planners, where do they come from, and what do they do.

To be very precise, product planning is supposed to deliver the following:

a. The product lines and product plan for each line.
b. The product-brief to the R&D.

Not very long ago, in most of the organizations, marketing function used to do this activity. In fact the modern definition of marketing was derived exactly from this activity. In ancient times, any industry was defined by a sum of two activities—"make" and "sell." "Make" was performed by production and "sell" by marketing. Later, this appeared too simplistic and people started seeing marketing role as creating the market and collecting the revenue. Even this was found insufficient and marketing started deciding the product as well. At this point, the definition of business itself was changed to—"chose the value, create the value and communicate the value." In this description, the first and last activity was conducted by marketing and the middle one by operations which meant R&D and production. Slowly, even this arrangement was found wanting as marketing appeared unable to balance the factors needed to create a product plan—the finances of the company, the operational competencies and capacities, and the market scenario. A neutral agency was needed to take a balanced view of these factors and thus the agency called product planning was born.

S. Chandra, *Aesthetics: Quantification and Deconstruction*,
Design Science and Innovation, https://doi.org/10.1007/978-981-10-6235-3_3

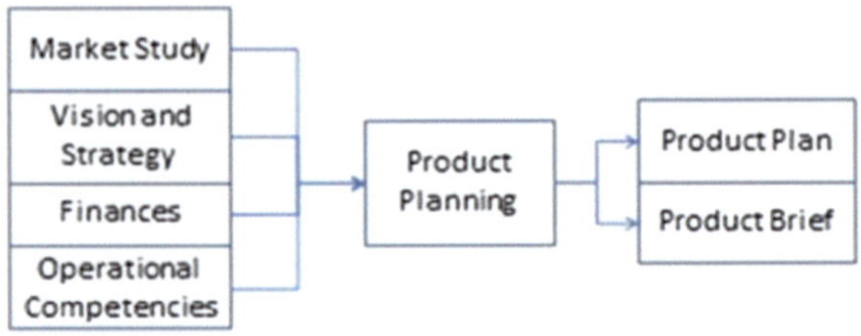

Fig. 3.1 Inputs and outputs for product planning

The diagram above (Fig. 3.1) summarizes the overall function of this agency. In short, based on the inputs like market scenario, the long-term and short-term vision of the organization, the financial overview, strategy and finally the operational competencies and capacity, the agency has to deliver the product plan and product-brief of each product. In real life, these two are not single point deliverables. Though ideally, everyone likes the idea of having a frozen product plan, a dynamic and agile product plan keeps changing with the ever-changing world. Similarly, the product-brief keeps evolving throughout the development cycle. Moreover, the product planner has to keep an eye on each stage to check whether the development is matching with the brief. Why? Why should it be so? Why cannot the designers deliver precisely matching with the brief at the end of the day? This is where the product planners' life becomes a challenge. The brief is never a mathematically precise document which can be delivered precisely. Therefore when the stylist comes out with the clay model, the product planner has to match it with his mental imagery, which itself is a bunch of vague concepts.

The purpose of this chapter is not to describe the methodology of product planning. What we are interested in is how a product planner finds it difficult to express what he wants from the design team and how he finds it even more difficult to vouch whether what has been designed is exactly what he wants. To answer the question, we have to consider what all he has to consider and what are the strategies before he actually delivers the plan?

3.1 DNA Strategy

DNA of design and design signatures are very commonly used jargons in the realm of design. All cars from BMW are recognizable by their front grill which is supposed to be their design signature. But the important question is—is this DNA of design an abstract thing like soul or a scientific entity? Looking at the characteristics of the biological term DNA and juxtaposing it against the world of artifacts—engineering, cultural, or otherwise—DNA of design appears to be a set of characteristics defining the products of a creator and making the creator instantly recognizable. Primarily, design is a process of wordless conversation, which the designer conducts with the user on one hand and the artifact itself on the other. Essentially, any conversation involves language and language, as Derrida says, is a complex artifact itself which needs to be deconstructed for its social, cultural, and psychological power equations. Now, in context of motorcycle design, this deconstruction involves decoding the language in which the user, the designer and the motorcycle converse with each other sending unspoken, subconscious messages, which are essentially a part of the DNA.

But the question which has been agitating us is the relationship between "DNA of design" and "design signature." Are they euphemisms for each other or are they mutually exclusive? The DNA can be defined as a code consisting of a set of characteristics, which makes the user of an artifact immediately recognize the creator. Going by the basic characteristics of a signature, it is, first and foremost, a visual entity and a part of the visual design, which, by definition, makes it a subset of the DNA. But what are common to both the terms are some characteristics:

1. Both the DNA and the signature are present in all products of a creator or a group of products representing a certain stream, whether it is an artiste or a company.
2. The characteristics, known as DNA or signature, are exclusive to the creator.
3. The characteristics should be user defined and not designer defined, i.e., the user should be able to identify and appreciate the characteristics.

Besides these common traits, the distinguishing feature between them is that design signature is essentially a sensorial feature whereas DNA is a set of properties encompassing all aspects. This makes signature an element of DNA.

Though, both the terms the DNA and the signature apply equally to creative art and design of engineering artifacts (in our case, motorcycles), we always believe that the semantic equation for design of engineering artifacts is more complex as the designer not only has to contend with semantic requirements but is also constrained by engineering factors and finally has to be governed by market dynamics. Since market is not a monolithic entity, the designers have to contend with varying tastes, economic profiles, demographics, and complex sociological equations. Fortunately, a creative artiste is free from these constraints in most cases (the most noted exception being cinema). Since this case study is for motorcycles, all these complexities have to be factored in the DNA of design, as the motorcycle is one of those rare technological species which is a cultural and engineering object at the same time.

Though, the DNA and the signature are a phenomenon common to works of art or commercial artifacts like motorcycle, the attitudes of creators toward their creations vary. Though in field of creative arts, creators are comfortable with signature, boardrooms and design studios for commercial products are sometimes apprehensive. Some common fears and apprehensions are as follows:

1. Design signatures or DNA are a constraint on flexibility and freedom of design.
2. For niche products, DNA is desirable. But for mass-market products, it is a liability.

Before we examine the truth behind these apprehensions, we can visit some areas of creative expression and verify these apprehensions.

Fine arts: Among European masters—Rembrandt, Van Gogh, and Salvador Dali the defining feature of Rembrandt is light and shadow play (Fig. 3.3), while Gogh (Fig. 3.2) is easily recognized by his flow lines. Dali's signature speaks through the surrealistic images in high-value primary colors.

Fig. 3.2 **Van Gogh**. *Note* the flow *lines*

Fig. 3.3 **Rembrandt**. *Note* the light and shadow play

Fig. 3.4 **Ramkinker Baij**. *Note* the figurative style in mud

Fig. 3.5 **Michelangelo**. *Note* the detailing of human anatomy

Ramkinker Baij and Michelangelo, two sculptors, separated by three centuries and a continental divide, belong to different cultural milieus. But the common trait with both of them is their strong signatures. Michelangelo's work can easily be recognized by the minute detailing of human anatomy (Fig. 3.5)—be it David's muscles and genitals or the ribs of Jesus in "Pieta." Ramkinker's signature is abstract but figurative style (Fig. 3.4) apparently in mud or cement.

Literature: Signatures are prominently available in literature as well. In fact, the impact a writer generates is directly linked to his signature. Hemingway has his hard metallic style which stands out in all his works whereas Shakespeare's signature is the internal conflict of the tragic hero beset with ambition and guilt of betrayal. Rushdie's imprint can be seen in his innovative vocabulary and

Table 3.1 Company DNAs

Company	Details
Hero (formerly Hero Honda)	Color-High value and high saturation. High unity for single-cylinder commuter bikes (2–3) and dynamism consistent with character (I for low cc and 3 for above 150 cc sporty bikes)
Royal Enfield	Minimalist bare construction for single-cylinder engine with single surface fuel tank, spoke wheel, and round head light. The gun logo and unique color, location, and shape combination for branding. High-vertical imbalance for Highest torque for a single-cylinder engine at lowest torque
Ducati	Red–black–white body color combination without stripes and with red colored frame. High vertical imbalance with high horizontal imbalance. Trellis frame with single side swingarm and L-twin engine arrangement
Harley Davidson	Multi-cylinder engine bare construction, single surface fuel tank, large wide tyres and round head light. Logo at head light center. Unique fluctuation pattern. High θ_1 and low θ_2. Large rake angle with high Rn. Belt final drive and V-twin engine arrangement. Very high torque at lowest rpm

Fig. 3.6 Bikes from makers with strong DNAs. Harley Davidson (*left* note the bare construction, single surface fuel tank and large rake), Ducati (*middle* note the *black*, *white* and *red color* combination, high-visual imbalance and trellis frame) and Hero Honda (*right* note the high value *high chroma colors* and high unity)

phraseology. The point here is that all the writers discussed above have traveled a huge range in their journey, still maintaining their distinct signature.

Cinema: Scorsese and Coppola have very clear stylistic signature of narration. Throughout his cinematic journey from "Mean Streets" to "Gangs of New York", Scorsese has his signature style-macho posturing, bloody violence, catholic guilt, redemption, and gritty locales, whereas Coppola has his sense of design in all his frames. Though both use the cinematic frame to create an epic like imagery, the selection of dominant colors by both is unique to both and creates the difference.

These examples essentially prove the point that DNA or signature does not inhibit the diversity of scope and the designer can always exercise his freedom of diversity maintaining his own signature or DNA. Another apprehension that DNA becomes a liability for mass-product companies does not pass the test of reality. We see that mass-product companies like Honda maintain different DNAs for different product lines and are able to maintain diversity while protecting the DNA. Table 3.1 describes the DNS of those companies (both niche makers and mass-market leaders) and we see that both types of companies have been able to maintain strong DNAs (see Fig. 3.6).

3.2 Architecture Strategy

Formally, architecture of a system is defined as "fundamental of a system in its environment embodied in its elements, relationships and in the principles of its design and evolution". A fundamental feature unique to consumer oriented engineering systems is its visual design, which is not present in the case of software systems. Thus, vehicle architecture can be defined as the aspect of design which characterizes the relationship between various elements of the design of a vehicle. If the quantitative aspect of a design fundamentally changes the relationship, it changes the architectural character in a significant way. To take an analogy from building architecture, if the number of pillars in a structure changes from 16 to 18, it is just a quantitative change. But if the number of domes on top of a structure changes from 1 to 4 it changes the architectural character. For a motorcycle designer, the concepts of architecture are to be applied to parameters of motorcycle design to find an objective formulation, which helps the designer in addressing the complexities of architectural change.

Fundamentally, in case of automobiles, vehicle architecture has two parts—engineering architecture and visual architecture. So far as visual architecture is concerned, it is in variably related to engineering architecture. Generally, with every ecological evolution, visual architecture takes a leap. One example is scooter another is step through motorcycles.

But why is it important to have an architecture strategy? We have seen through the history that history has been created whenever major architectural change takes place. The tectonic shift in Indian motorcycle industry happened when Hero Honda introduced four stroke engines in a market cluttered with two stroke engines. At the same time, architectural changes are difficult and risky. Again, when Hero Honda in India introduced step-through motorcycles in India—remember, it was step through which wrote the success story of Honda through model D and model E—it failed miserably. This change was done at a huge cost, a whole new plant was erected for it but the Indian sensibility was not prepared for it. So, architectural strategy, in many ways, is the backbone of product strategy and should be made with great care.

3.3 Platform Strategy

The purpose of platform strategy is to create common bases for engine frame to generate various combinations by adding differentiating enablers. Platforms are these large common bases predominantly in form of frame (for motorcycles)/ chassis (four-wheelers) and engines. The biggest challenge for a product plant from point of view of infrastructure is to create maximum amount of differentiation in product by using maximum amount of common parts. The answer to this challenge is to have a limited number of frame and engine platforms and create as many combinations with new body designs and features as possible.

Different makers have strategized their platforms differently. Bajaj Auto has devised its platforms on product families. All Pulsars have a common frame (Well! Almost.), combining them with different sized engines to create different Pulsars. On the other hand, Hero has strategized their platforms on engines sizes. All 100 cc bikes have almost the same frames and same engines combining them with different bodies and features.

3.4 Product Family Strategy

Both the architecture and platform strategies are designer centric. But the real center of product planner's universe is the user. How does the user look at the products of a company? To start with, manufacturers have their product lines and each product line has segments based on the buying capability of the users or price points. So we have entry segment, executive segment, premium segment and luxury segment. Within a segment there are users having different choices and how do we expect the users to know that we know their choices? This is where product families play their role. Moreover, families can exist across the segments. So, Bajaj Auto has Discover family and 'Pulsar' family. Hero has Splendor family, Passion family, and Karizma family.

What is the basis of creating families? Fundamentally, it is the user's perception about a product. There are two types of perception-desirable, which the users want to persist with and undesirable which the users want to do away with. So evolution continuing with desirable perceptions in a family becomes inevitable. So we have Royal Enfield continuing with Bullet with so many variations of it.

3.5 Semantic Strategy

Semantics means the science of meaning. An automobile user communicates with his vehicle at two levels of meaning:

1. The functional meanings like affordances (what all, the user can do with the machine and how well), visual metaphors, and categories. Krippendorf calls them meanings in use.
2. The reflective meanings where the user interacts with the machine at an emotional level like character, verbal metaphors, etc. Krippendorf calls them meanings in language.

The product planner's plan has to describe, categorize, and strategize these meanings into his DNA strategy, architecture strategy and product family strategy. In fact the product family strategy has to largely depend on semantics. So far as the functional meanings are concerned there are tools to quantitatively qualify them. The problem arises when the meanings in language have to be strategized because there are no quantitative tools and it largely depends on subjective interpretations.

3.6 Product Planning Tools

I am not discussing the tools in detail here. The point of discussion, in fact, is to see whether these tools help the product planner or strategist to take decisions without making subjective evaluations or interpretation. First of all, I have not come across any tool, which deals with aesthetics exclusively. In fact, even Krippendorf, while discussing meaning in language talks in qualitative terms only. In fact, most of the tools like Kano's model or weighted scoring or buy the feature are meant to evaluate features based on subjective ratings. Different organizations use different tools but these tools can be classified in three categories:

a. **Market Research**: One of the most used tools is conjoint analysis. This helps the planner to decide the value people assign to features, benefits, and attributes individually and in combination. This is a survey based statistical tool which determines the effect of price changes on people's choices and the effect of one attribute on another. Let us understand like this. A planner has seven features to choose from and these features add up to a price increase of $140, whereas he has a budget of $50 only. The planner will conduct a survey and will decide with the help of conjoint analysis the combination of features adding the most value to the product.
b. **Design Positioning**: This tool helps the planner to decide the engineering and aesthetic positioning. The prevalent tool for this is bubble diagram. All competitors are placed on a graph according to their engineering and styling positions and the planner tries to find apposition for the new product on this diagram.
c. **Briefing**: This is a document with which the planner briefs the design team about the expectation from it. It defines the price range, the customer segment and the engineering and styling requirements. The problem for the planner arises exactly at this point. Though he can precisely define the product family, platform, and the engineering architecture, he gets stuck with words and phraseologies when he comes to visual architecture and semantic requirements.

3.7 The Tensions

The whole duration of product development is filled with tensions. The product planner comes in conflict either with the designer or the marketer or the project leader. The tensions can be categorized as:

a. Utilitarian versus hedonic benefits: The most ancient tension for automobile designers has been between vehicle performance and fuel efficiency and to strike the right balance has been giving headaches to the planners. But this is a debate between hardcore thermodynamic engineers. The more typical frictions are between stylists and engineers where engineers insist on change in styling of lighting surfaces to get adequate illumination or seat surfaces for ergonomics or body profile for aerodynamic performance.
b. Creativity versus schedule, process, and cost: This is the conflict which the planners have to face almost everyday. What compounds the complexity of the situation is the fact that aesthetics in nonquantifiable and there is no way that one can reach to an optimal solution mathematically.
c. Planned aesthetics versus actual aesthetics: The product planner has to fundamentally understand the emotion needs of the user and communicate it to the designer. The language and the grammar of this communication is, at best, vague, and subjective. The meanings of words are derived by the individual who receives it and it is not necessary that they will be same as that intended by the communicator. To put it simply, this lack of a scientific language and grammar of aesthetics results in communication gaps between the planner and the designer. As a result, when the designer comes out with his design in form of sketches or clay model, invariably the planner finds it not matching with what he has in mind. This results in a continuous tension and sometimes it is not resolved even after the product goes to market.

In fact, product planning is all about communication. And the communication process is built around weaving a narrative and creating a web of emotions around this narrative. The narrative can be about a Himalayan journey on rough and hostile roads encountering difficult seasons and the web of emotions can be adventure, dynamism, grittiness, and courage (see Fig. 3.7). The narrative of Che and Alberto on a Norton 500 across the South American continent has already become a legend. Similarly, the narrative of romance and gang-wars around motorcycles (Harley and Triumph) has been immortalized through "The Wild One." Again this narrative is built around the emotions of glamor, romance, and rebellion. The whole engineering structure is built around this combination of narrative and emotions. The engine performance has to match this narrative. The motorcycle dynamics has to match this narrative. Finally, the styling has to match this narrative. The problem lies in communicating this combination of narrative and emotions.

The triangle of user, artifact, and designer has the product planner at the center of gravity. If the designer has to have a second order understanding of the artifact (an understanding of user's understanding of the product), the planner has to have both

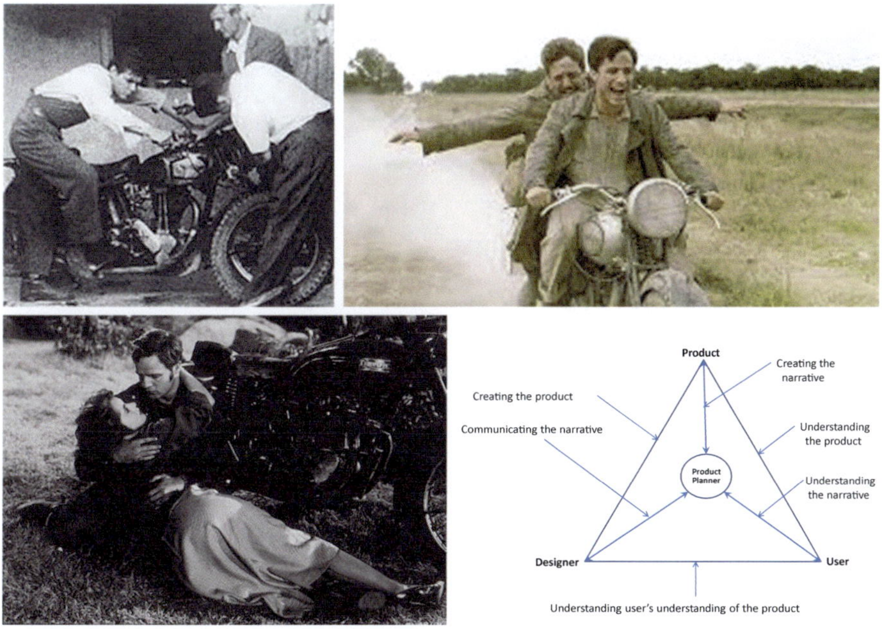

Fig. 3.7 Creating the narrative

the second and third order understanding. It is he, who makes available the user's understanding to the designer. Most importantly he has to understand the designer's understanding of the user's understanding of the artifact. Now this complex network of understandings is relatively manageable, when it comes to engineering because the vocabulary and the science of numbers are readily available and the communication gap is zero. If the product planner wants an acceleration of x m/sec^2, he can communicate it precisely and the engineering designer can very precisely agree or disagree to deliver it. Finally, after the product is developed, it is precisely verifiable as well. Alas, the same is not true for aesthetics. So, at all levels of strategizing aesthetics becomes the Achill's heel. Whether the planner needs to decide the DNA of product line or the visual architecture or defining a product family in terms of aesthetics, it all becomes a play of words to be interpreted differently by everyone. Moreover, the question of degree becomes impossible to communicate. Similarly, the tools of product planning deal very effectively where the parameters and factors can be expressed in terms of numbers. In case of aesthetics, they too become helpless. Tragically this is the world of helplessness, which he has to deal with everyday. If I am in his place, I will start everyday looking up to the sky and praying for a science, which can specify, quantify and qualify aesthetics for me. The next chapters of the book are an attempt to answer this prayer.

Chapter 4
Challenges for Styling Designers

All of us know Pablo Picasso, Salvador Dali and Michelangelo and their creations. And if someone has not heard about them, we feel entitled to look at him with amusement. I wish we could feel equally entitled to look with amusement at someone who has not heard who designed the iconic BMW R32 or Harley V-Rod. Or, for that matter, who designs the Ferraris and Pininfarinas. One reason, of course, is that Picassos and Dalis are freelancers whereas the automobile designers are not. They have to sweat it out in design studios of manufacturers without anyone coming to know about them. But the point is what really the difference between the two is? The artistes are free from commercial considerations—though their creations fetch millions. But for the stylist, it takes everything that it takes for an artiste. Additionally, he has to cater for engineering and cost. But most importantly, he has to sell a product, which should cater to the emotional needs of all users. In short, he has to do something incredibly more complex than an artiste and still achieve what an artiste achieves.

4.1 The Process

My purpose here is not to sing paeans for the stylist here. The basic objective is to understand what exactly he (or she) does and the challenges he (or she) faces. You will get no marks for guessing that the main function of a stylist is to design the appearance of an artifact. As we have discussed in the first chapter on semantics, this involves creating meaning in use, meaning in language and meaning in ecology for the desired artifact. If all this sounds too complicated, the fact is, this *is* complicated. But, before we discuss the reasons, let me explain the usual sequence of events for a stylist (see Fig. 4.1).

We have discussed in the previous chapter the challenges faced by the product planners in formulating an objective product brief. We can see here that the challenges faced by the styling designers start from the very start. But, this we will

S. Chandra, *Aesthetics: Quantification and Deconstruction*,
Design Science and Innovation, https://doi.org/10.1007/978-981-10-6235-3_4

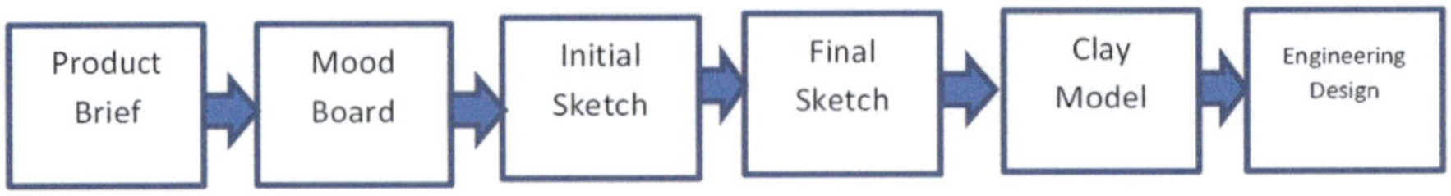

Fig. 4.1 Sequence of events for automobile styling

Fig. 4.2 Example for a mood board

discuss later, after explaining the process. In short, what the stylist receives from the product planner are a very vague idea about the emotional needs of the targeted customer, the profile of the customer himself and the engineering specifications.

The most important step for a stylist is the next step of creating the mood board (see an example in Fig. 4.2). This is the step where, the stylist has to translate the vague emotion-brief into design characteristics, to show to the prospective customer —in this case, the product planning which confirms whether this board matches with their vague characterization. This depends on the deep understanding the designer has about the relationship between design and emotions. The problem is this deep understanding itself is instinctive, which in most cases takes away the objectivity.

Once a consensus is achieved on mood board, the designer starts working (Fig. 4.3, top) on the product. This needs inputs from engineering team as well like ergonomics, geometry, and manufacturability, and so on. But the chief output at this stage is still some free-hand lines drawn with a pencil. At this stage, the stylist works with many ideas and continues on creating new ones till it hits the bull's eye. The next step is firmed-up sketches (Fig. 4.3, bottom), which means realistic images which present the product as the user is supposed to see it. This includes the color-scheme, the engineering features, and geometry. This is a complicated job

Fig. 4.3 Free-hand (*top*) and firmed-up (*bottom*) sketches

which needs inputs from and consensus with all stakeholders. Most often nowadays, this sketch is in 2D but with advent and advance of digital technology, people have started advocating for sketches directly in 3D and softwares are readily available for this.

The story for the stylist reaches its climax with clay model (Fig. 4.4), which is nothing but a hard visualization of the image. Today there are many routes available to reach at this clay model. The traditional route is one where an expert clay-model shapes a lump of malleable clay or any other such-substance to match with the firmed-up sketch. In case of simpler objects, it is not that complicated but for automobiles it is tough as accuracy of geometry, symmetry of surfaces, and quality of surfaces become important criterions. But the method currently popular is through clay milling, where the sketch is first translated into surfaces in computer and then clay (or wood) is milled with the surface as the input data. This simplifies many problems present in the traditional method but traditionalists abhor this method. This is because they believe in traditional power of mind, eye, and hand coordination available in shaping clay with hand. But like everything traditional, the manual clay is today counting its last days. One of the chief reasons why the digital method is preferred besides the simplicity and accuracy is that the surface becomes available for engineering analysis even before the clay is ready. This results in iterations in clay if engineering analysis demands and by the time the management approves the clay there are very few possibilities of further iterations. On the other hand, in traditional method, the clay, after the management approves it, has to be scanned and translated into a surface. This means, any iteration after the engineering analysis comes only after the management has approved it. This naturally causes many difficulties which everyone wants to avoid.

Fig. 4.4 Car clay model

4.2 The Tensions

Amidst all the hard work that a stylist has to put in during this cycle, the central theme that is running throughout is emotion. In the very first thing that he (or she) has to do is to convert the brief into a mood board. Here, as we have already observed, the brief is very vague and sketchy. For example, sample this portion from a product brief:

Product aspirations:

a. *Unique and rugged yet acceptable in formal settings*
b. *Sturdy, bold, powerful yet premium*
c. *More formal but exclusive*
d. *Showcases success and maturity*

Lifestyle aspirations:

a. *Clean, superior and relaxed*
b. *Explore new places and things*

Target consumer:

a. *Around 20 years old undergraduate student*
b. *Explorer*

c. *Aspirer*
d. *Exclusivity seeker*
e. *Needs powerful mobility to take him anywhere without boundaries*
f. *Agile*

Concept statement:

Explorer: Escape the ordinary, lead the pack from Ladakh to cyber city, and take it anywhere

Product attributes:

a. *Upright and confident*
b. *Rugged yet refined*
c. *Stand out*
d. *Agile*
e. *Powerful*
f. *Sharp eyed*

What do we make out from this? This product brief expects the stylist to be sociologist, psychologist, artiste, and an engineer put together. To cut the long story short, will it not be more useful and effective if the stylist is given a mix of desired emotions in the desired proportion straightaway? But, there is no such scale available to specify and quantify these emotions and the stylist has to depend on his own instincts and understanding of emotions and cultural sensibilities. And therefore, the product brief becomes the proverbial elephant as perceived by seven blind men and each of the blind man perceives the elephant based on the organ that he touches.

But the difficulties of the stylist do not end here. Mood-board demands one more obstacle to be crossed. And this is converting the emotion mix to geometric and engineering parameters like shape characteristics, space combinations, unity, balance, color, and so on. Currently, there is no available grammar to do this and each stylist is left to his (or her) own devices. Sure, absence of grammar multiplies the creative possibilities but then this is like leaving the artiste in an ocean of possibilities without even a little boat to sail through.

Now, when it comes to the sketch, the poor soul finds that he (or she) is in a deeper darker ditch. Here, the styling parameters have to be driven down to every visible element of the artifact. For example, in case of motorcycles, the fuel tank, and the seat have to be assigned their shapes, colors and their mutual relationship. Moreover, the overall visual architecture of the bike has to be decided. All this needs to be derived from the emotion mix and there is no relationship available between the characteristics and the emotion mix. Again, thc stylist is left with his (or her) own understanding.

The toughest battle for a stylist is the presentation of the clay model to the management team where the management is to be satisfied that the product brief is entirely met by the clay model. Here the complexity of the task gets doubled as one set of blind men has to convince another set of blind men that the elephant that they have created matches exactly with the expectation of the other set. Now compare this task to the task of an engine designer who has been asked to design an engine with a certain set of power, torque, fuel consumption, noise, vibration, and acceleration. The engine designer has his (or her) task cutout. The actual values delivered by the engine have to match with the set of expectation and his job is done. No such luck for the stylist.

Today, styling has evolved a lot. But most of the evolutions are in field of technology. Sketching can be done directly in 3D. Surfaces can be created directly. Clay model can be generated after doing all sorts of engineering validation and with utmost accuracy. What is not available is a tool-kit for the stylist to understand the aesthetics, express it objectively and quantitatively, drive it down to individual parts and characteristics and finally measure the emotions. This is the void, this book attempts to fill.

Chapter 5
The Science of Emotions

> How often people speak of art and science as though they were two entirely different things, with no interconnection. An artist is emotional, they think, and uses only his intuition; he sees all at once and has no need of reason. A scientist is cold, they think, and uses only his reason: he argues carefully step by step and needs no imagination. That is all wrong. A true artist is quite rational as well as imaginative and knows what he is doing. If he does not, his art suffers. The true scientist is quite imaginative as well as rational and sometimes leaps to solutions where reason can follow only slowly; if he does not his science suffers.
>
> —Isaac Asimov, "Promethius" The Roving Eye (1983)

While reading this quote, I was wondering how aptly it captures the field of automobile design, where art and science live inseparably and intertwined with each other. The artist has to think of aerodynamics and the engineer has to think of aesthetics. And then comes the question—what is the science of aesthetics and is there a science of aesthetics at all? And then the question arises—what is the science of emotions and is there a science of emotions at all? And then strikes the question—what is science after all?

> However, all scientific statements and laws have one characteristic in common: they are "true or false" (adequate or inadequate). Roughly speaking, our reaction to them is "yes" or "no". The scientific way of thinking has a further characteristic. The concepts which it uses to build up its coherent systems are not expressing emotions. For the scientists (there is) only "being" but no wishing, no good, no evil, no goal. As long as we remain within the realm of science proper, we can never meet with the sentence of the type "Thou shalt not lie." There is something like a puritan's restraint in the scientist who seeks truth: he keeps away from everything voluntaristic or emotional.
>
> —Albert Einstein, Essays in Physics (1950)

So, exploring the science of emotions means defining, classifying, and quantifying emotions without being emotional and judgmental. It is not something which has not been tried in past. Pythagoras, the ancient Greek philosopher, musician and mathematician famously said—"everything is a number" and proved it by establishing relationship between emotions and octaves in music by his discovery that musical notes can be translated into mathematical equation and thus mathematical

S. Chandra, *Aesthetics: Quantification and Deconstruction*,
Design Science and Innovation, https://doi.org/10.1007/978-981-10-6235-3_5

equations decide the length of strings creating different emotions through octaves. In a way, we are trying to explore through this study for motorcycle design what Pythagoras did for design of music and to do that, we need to objectify and quantify emotion itself and to that we need to define emotion as a scientific entity.

5.1 Emotion Perspectives

To start with, we can define emotion as the response of human being to an artifact on his interaction with the artifact. This artifact can be a hard one like motorcycle, a car or a building or a soft one like an incident, a story, a poem, a song, a movie, or a rainbow in the sky. But, here we are talking in context of hard objects, which the designer wishes to interface with the user. The purpose of this interface is to make the artifact usable. The interface results in stimuli and stimulus congruence and facilitates impression formation which finally affects consumer responses [1]. Now, in how many ways does a person interact with the artifact? According to Norman [2], a human being perceives an object at three levels:

1. Visceral level, at which the object appeals to five human senses.
2. Behavioral level, at which emotions are generated due to the functional behavior of the object.
3. Reflective level, which appeals to the human brain conditioned by cultural experiences and thinking process.

The ecological evolution of design can be understood only with the help of tools of semantic approach. In fact, the approach to design itself has evolved over time. Though the semantic approach came in vogue only in 1987 with the publication of "The Semantic Turn" by Krippendorff [3], the evolution itself has been largely semantic since the inception of automobiles. While talking of meanings of an artifact, as seen by designer and user, Krippendorff categorizes them as meaning in use, meaning in language and meaning in ecology. While considering the meaning of motorcycles in use as seen by the user, it amounts to (1) the interface of rider with the motorcycle (or ergonomics), (2) the visual metaphors (the visual scheme or architecture), and (3) the affordances of the motorcycle (the various things that the motorcycle can do for the user or the user can do with the motorcycle) as a whole and in parts. More important in this context is second-order understanding of the motorcycle (the understanding of the designer of the user's understanding of the artifact) as a whole, in parts and in parts constituting the whole. According to Ross and Wensveen [4], aesthetic interaction is an interaction experienced with a product or a system that (1) has practical use next to intrinsic value (2) has social and ethical dimensions (3) has satisfying dynamic form and (4) involves the whole human being. So, with time emotional design has come to occupy the leading role in new product development. The traditional approach to design takes into account the functionality, the durability, the aesthetics, the manufacturing process and the cost into account whereas the modern approach goes into history, geography, psychology, sociology, and politics (yes, politics! as you relate to the interrelationship between the social and economic forces in society affecting the design) of the artifact besides the engineering, whereas Chitturi et al. [5] define design as a trade-off between emotional and utilitarian benefits and preventive and promotional emotions. In context of a motorcycle this applies in an even more complexity.

Emotions are as old as humanity and humans have tried to interpret it scientifically and mathematically since the beginning. Pythagoras in Europe and Bharat in India are few ancient examples. In modern times, Plutchik [6] (Fig. 5.1) has designed a three-dimensional model to describe emotions, on the pattern of a color wheel, where coordinates r, θ represents the range of emotions and the third vertical dimension represents the shade of emotions. This model has the extreme shade of various emotions on the top circle and as we move downwards, we find lighter

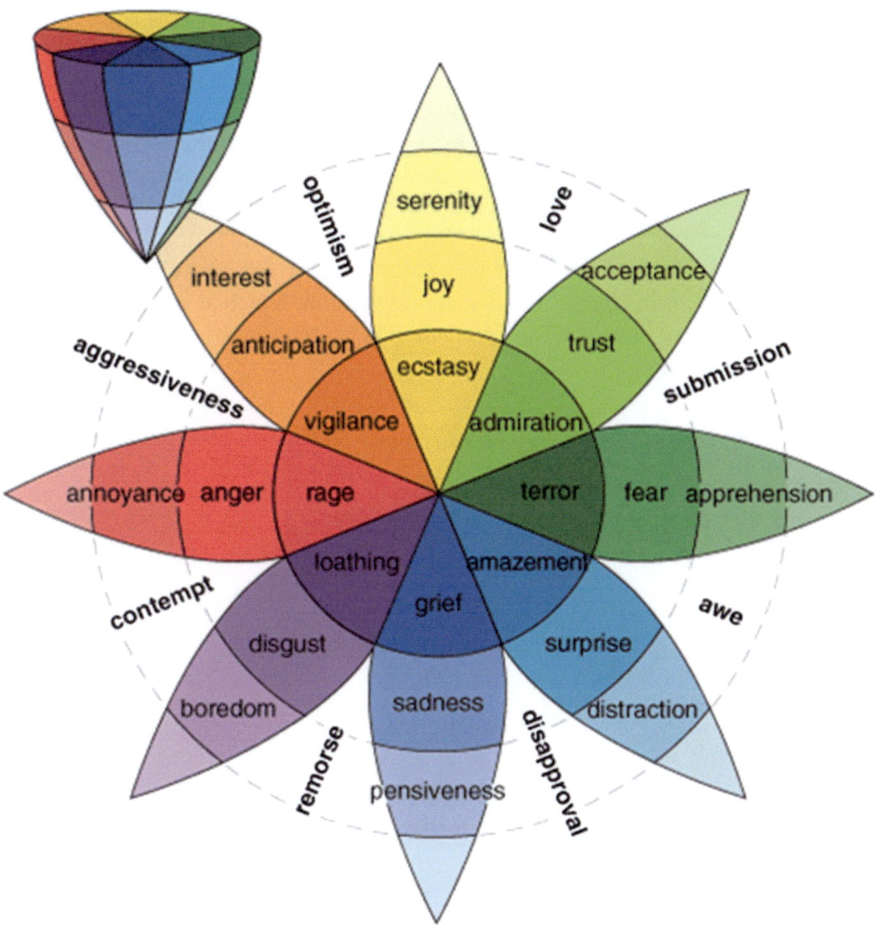

Fig. 5.1 Dimensional model for emotions by Plutchik

shades of the emotion. For example, we have grief on the top circle and moving downwards, we have sadness and pensiveness in that order. So, on the top circle we have the emotion and its intensity and on the third axis we have the shades. Another aspect of this model is the dividing line for each hue of emotion. Between the hue of grief and amazement, we have the dividing line of disapproval. Similarly, between the hue of amazement and terror, there is the dividing line of awe.

Vitters et al. [7] have discussed various models to accurately plot and quantify emotions and their respective strengths and weaknesses. They argue that the neurological model by Panksepp [8], where human emotions can be described as seven basic traits, i.e., fear, anger, sorrow, anticipatory eagerness, sexual lust, and maternal nurturance is neurologically adequate as it is coherent with the brain system but psychologically inadequate. Vitters et al. then argue that the model offered by Communicative Theory of Emotions by Oatley and Johnson-Laird [9] gives a suitable perspective of basic emotions. The authors finally argue that besides neurological and psychological perspectives a third socially oriented perspective is required, which, provided by Russel and Barret model is known as Circumplex of Emotions. This model geometrically defines emotions in terms of two axes, i.e., deactivation to activation and unpleasant to pleasant.

From the above discussion, we conclude that though the Plutchik model is most accurate to measure the exact hue and intensity of emotion, it is too complex for our purpose. The neurological and psychological models are mathematically inadequate hence the Russel and Barret model is most appropriate. This model is, in fact an interpretation of Watson and Telegen's circumplex of emotion (Fig. 5.2), where emotion has two coordinates—valence and activation. Valence in simple terms is extent of pleasure (or lack of it) and activation is the amount of excitement (or lack of it) an artifact generates.

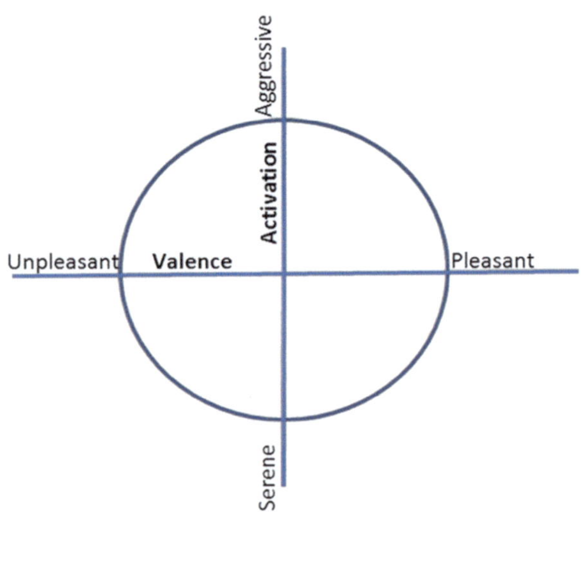

Fig. 5.2 Russel's interpretation of Watson and Tellegen's circumplex of emotions

The diagram shown here in Cartesian coordinates can be expressed in polar coordinates as well, where the type of emotion can be expressed in terms of angle "θ with reference to X axis and the intensity of the emotion can be expressed in terms of radius "r". Interestingly, this theory by Watson and Tellegen finds a remarkable convergence with the grammar of ancient Indian literature. While going through "Rasa-Siddhant" by Bharat, we come across the theory of generation of rasa (i.e., emotion), (Barlinge [10]).

Vibhavanubhava vyabhichari sanyogat\rasa nishpattih

This defines three elements for generation of emotions:

a. *Vibhavanubhava*: actions and situations generating emotions
b. *vyabhichari*: communication of emotions
c. *sanyogat*: the first two elements combining with permanent emotions residing in the recipient.

This means actions and situations stimulating an emotion are communicated to the recipient, where the codes for various emotions get activated after receiving the communication.

Further as Barlinge quotes from "Rasa-Siddhant," Bharat defines the points of communication process for any art form:

a. *Kalaakar* i.e., The artist
b. *vastunishtha* i.e., the creation or the work of art
c. *sahridaya* i.e., the recipient

This perfectly matches with the design process defined by Krippendorf, who identifies the three points of this process as:

a. The designer
b. The artifact
c. The user

But the most remarkable convergence is observed between Bharat and Watson and Tellegen, where Bharat defines "siddhi" as basic element of "rasa" having two types, i.e., (a) "manushi siddhi," (b) "daiviki siddhi." The first one has two extremes—*harsha* (Joy) and *vishaada* (Sadness). The second one also has two extremes—*bhavaatireka* (extreme excitement) and *atmaleena* (extreme serenity). These can be compared with the two axes of the circumplex of emotions—valence and activation.

Bharat, in this study of emotions, has enumerated nine rasas, which cover the whole gamut of human emotions:

1. Shringaara: Love, beauty, devotion
2. Shaant: Peace, relaxation, calmness
3. Karuna: Sadness, compassion, pity
4. Bhayaanak: Fear, anxiety, worry
5. Vibhatsa: Disgust, depression

6. Raudra: Anger, irritation, violence
7. Vira: Courage, confidence, pride, aggression
8. Haasya: Joy, humor, sarcasm
9. Adbhuta: Wonder, mystery, curiosity

Having established the convergence, the next point is to locate the emotions on the circumplex which means it is necessary to relatively quantify the elements of valence and activation for each emotion. For this purpose two types of tools have been suggested in various studies: (1) Psycho-physiological tools and (2) Constructivist psychological tools.

(a) Psycho-physiological tools: Levy [11] categorizes these tools based on types of physiological responses:

   a. Central nervous responses through Electroencephalograms (EEG) or Functional Magnetic response Imagery (fMRI) or near-infrared spectroscopy.
   b. Peripheral responses like heartbeats using electrocardiograms (ECG) or body temperature through infrared thermography.
   c. Motor responses like eye movements and pupil dilatation through electromyography: Rauterberg [12] also talks about EEG and fMRI to monitor brain activity variations but prefers electromyography as more practical.

Wollmer et al. [13] in their paper introduces a new technique for continuous emotion recognition in a 3D space spanned by activation, valence and time using long–short term memory recurrent neural networks (LSTM-RNN) finds it better than another technique called support-vector-regression (SVR) as the former gives results matching to human perception.

Dzogang et al. [14] mapped the central nervous responses conducting two different exercises. In first one, two songs "Ode to Joy" that was full of love and happiness, and a sad song "You are not Alone" were used for mapping valence and activation. In another exercise, a movie "Little Miss Sunshine" was fragmented into 24 sequences with different emotions and each sequence was mapped for valence and activation. We translated the results in following table on a scale of 1–10 [see Table 5.1 (for the first exercise) for songs and Table 5.2 (for the second exercise)].

(b) Constructivist psychological tools: For this, we conducted a small experiment with college professors specializing in Hindi literature, well versed in rasa-siddhant and with intricacies of poetry. They were given poems

Table 5.1 Dzogang's valence and activation mapping for songs

Song	Emotion	Score	
		Valence	Activation
Ode to joy	Love, happiness	10	5
You are not alone	Sorrow	4	2

Table 5.2 Dzogang's valence and activation mapping for movie sequences

Sequence	Emotion	Score	
		Valence	Activation
3	Anger	4	8
5	Anger	4	9
7	Courage	5	10
8	Anger	4	7
9	Sorrow	6	2
11	Revulsion	3	7
13	Love	10	5
14	Wonder	7	6
15	Anger	4	6
16	Anger	4	8
18	Ugliness, hatred	1	5
20	Humor	8	3
23	Introspection, quiet	4	1

The sequences 1, 2, 4, etc., have not been discussed in the paper

representing different emotions and were asked to rate the elements of Aanand (pleasure)] uttejanaa (excitement) on a scale of 1–10. The results were tabulated as follows (Table 5.3):

Now if we superimpose Bharata's rasa-siddhant on to Russel's emotion wheel using the above results, the nine rasas are mapped on the two coordinates of valence and activation as shown (Fig. 5.3).

Here, each of the emotion is expressed by a specific angle, determined by the degree of valence and activation. The angles can be tabulated as (Table 5.4):

Moreover an arbitrary scale from 0 to 3 has been generated for the intensity of the emotion where the complete absence of the emotion can be expressed by 0 and the ultimate extreme is expressed by 3. Now, from the designer's perspective, it can be safely concluded that no designer designs a motorcycle with the emotions

Table 5.3 Estimation of valence and activation for poems

Poem	Emotion (Rasa)	Score	
		Valence	Activation
Main neer bhari dukh ki badli- Mahadevi Verma	Karun	3	2
Martya manav ki vijay ka turya hun main- Dinkar	Veer	4	8
Beeti vibhavari jaag ri- Jayshankar Prasad	Shaant	5	1
Anarky- Dinkar	Haasya	8	2
Kah do Shankar se aaj karen- Dinkar	Raudra	3	8
Prabhu tum kar do vaman- Nagarjuna	Vibhatsa	2	6
Dheera sameere yamuna teere- Jaidev	Shringaara	10	5
Lanka burning episode in Ramacharitamanasa- Tulsidas	Adbhuta	8	8

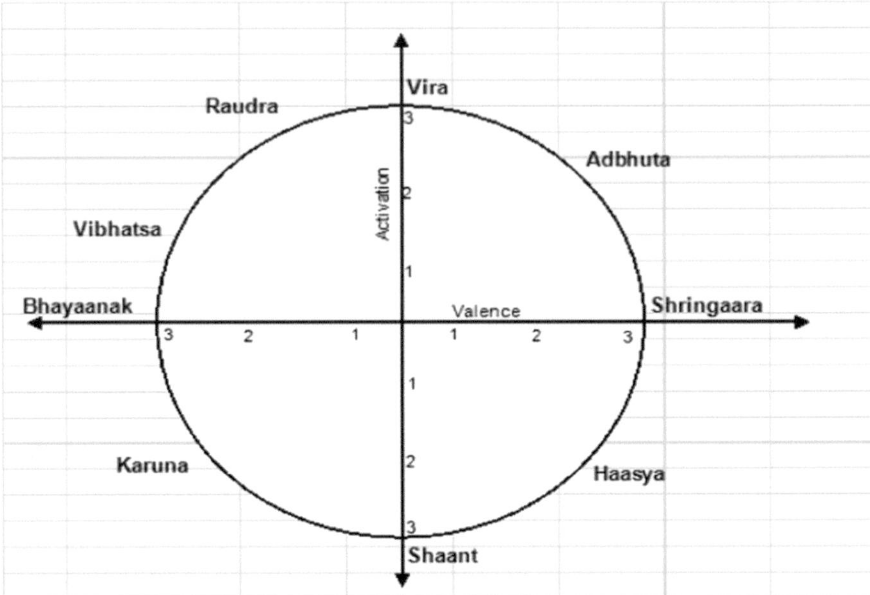

Fig. 5.3 Rasas on Russel's emotion wheel

Table 5.4 Angular coordinates for emotions

Emotion	θ (°)
Shringaara	0
Adbhuta	45
Vira	90
Raudra	120
Vibhatsa	150
Bhayanak	180
Karuna	225
Shaant	270
Haasya	315

"Bhayanak," "Vibhatsa," "Karuna," and "Haasya." Moreover, the emotion "Adbhuta" relates to the element of wonder, which in essentially the result of novelty of other emotions. So this emotion is a derivative of other emotion. So we are now left with the emotions of Shaant (which can be translated as "Sober" in context of motorcycle, Shringaara (Glamor), Vira (Dynamism), and Raudra (Intimidation). So the emotion generated by any automobile or motorcycle can be expressed as a set of four numbers *S* (Sober), *G* (Glamorous), *D* (Dynamic), and *I* (intimidating), where each number represents the intensity of respective emotions on a scale 0–3.

Having established the basis for quantification of emotions, another challenge remains for an artifact designer—that of relating the emotional coordinates to the market value, he seeks to generate. This is an extremely complex area due to the vagaries of subjectivity, cultural, geographic and demographic variations. Prinz [15] identifies three elements here which are different but related—emotion, appreciation, and aesthetic value. They are related as emotion plays a role in directing our aesthetic preferences and appreciation is an emotional state. But, ironically, there can be situation where one can appreciate without being emotional and conversely, emotions can be generated without appreciation. Lee et al. [16] add a fourth dimension to this three-dimensional paradigm—that of culture. Now, any artifact used by a user is a cultural object, which is defined by the authors as:

> Cultural products are the final outcome, produced in the process of industrial activity using cultural materials. It is an idea-intensive and technology-intensive undertaking to produce high value added products. As a product of culture, technology, and creation, it has the characteristics of multiplicity, diversity, and high-added value.

The authors then classify the cultural products as endogenous (which use the cultural identities) and exogenous (which generate their own cultural identities, e.g., Harley Davidson, which has become an icon of American culture).

Now this four-dimensional paradigm of value, emotion, appreciation, and culture gives us an insight into the complex web of relationships leading to the artifact's aesthetic value in marketplace. Among those, who have tried to translate these relationships into mathematical algorithms, Carter [17] has suggested that various stakeholders in the product have to be taken on board and their responses have to be summed up giving due weightage. Schummer et al. [18] enumerate three basic elements for determination of aesthetic value—symmetry, mathematical simplicity, and style of experimentation. But they don't provide a mathematical model. Jones et al. [19] have indeed given an equation for visual quality, but in context of a township

$$\mathrm{VQ} = (I + V + Q)/3, \tag{1}$$

where VQ = Visual quality, I = Intactness, V = Vividness, U = Unity.

They further quantify the change in visual quality (before and after) as

$$R = (\mathrm{VQ}_{\mathrm{Final}} - \mathrm{VQ}_{\mathrm{Initial}})/\mathrm{VQ}_{\mathrm{Initial}}, \tag{2}$$

where

R Ratio of change in visual quality.

The visual impact measured at given viewpoints is

$$\text{Visual Impact} = R\,.\,P, \tag{3}$$

where

P Population viewer contacts per year.

The problem with this methodology in the context of motorcycles and the scope of our study is twofold:

1. It considers visual quality completely unrelated to emotions.
2. Market segmentations of motorcycles are not taken into considerations. This is due to the fact that this method applies the same yardstick to all motorcycles.

A significant study was conducted by Grimm [20] for quantification of aesthetic value in context of wind farm landscape. Here, the author has formulated the following quasi-static parameters:

1. Visual Landscape Character Value (VCLV): Here the whole landscape is looked from various viewpoints and assessed in terms of:

   a. Relief: Complexity of topographic variation
   b. Vegetation Cover: Vegetation extent and potential to screen and cover views
   c. Infrastructure and built form
   d. Culture landscape

Each parameter is defined on an objective scale of 0–9 and overall character is summed up both as percentage and out of 20 as an aggregate for each viewpoint.

2. Detailed Visual Effect of Development Form (VEDF): Here also the visual effect of the development (the differential of with and without the development) is assessed in terms of:
1. Horizontal Visual Effect
2. Vertical Visual Effect
3. Distance visual effect

Each effect is calculated in terms of with and without and differential percentage is averaged.

3. Percentage of Visual Change (PVC): This is the coefficient of visual change to the baseline VCLV for each viewpoint, i.e.,

$$\text{PVC} = \text{VEDF (as percentage)} \,.\, \text{VCLV (expressed as value out of 20)}.$$

Though this method is quite objective and least influenced by subjectivity, here also we face the same problem as that for Jones method, i.e.—(1) No relation to emotions and (2) No relation to market segmentation.

So the studies conducted so far does not fit in case of motorcycles where value is to be derived from emotions based on market segmentation. Since, we can safely say that the aesthetic value of an automobile is a function of the intensity of various emotions generated by it, it can be mathematically denoted as:

$$\begin{aligned} V_{\text{aesthetic}} &= f(r, \theta). \\ &= f[r]_{s,g,d,i} \end{aligned} \tag{4}$$

Here, we have to make an assumption, that before the design process starts, the marketers define $[r]_{s,g,d,i}$ to ensure that the product is able to generate the desired emotions leading to desired appreciation leading to desired value proposition for the customer. Now if the designer is able to generate exactly the same emotional coordinates through his creation, he is able to generate the aesthetic value of 1 which we decide to be maximum value on the scale. Deciding 0 to be minimum value on the same scale, we can mathematically decide the relationship as

$$V_{\text{aesthetic}} = 1 - \left(\left[\sum \delta r \right]_{s,g,d,i} \Big/ \left[\sum r_{\text{desired}} \right]_{s,g,d,i} \right), \tag{5}$$

where

$$\begin{aligned} \delta r &= [|r_{\text{desired}} - r_{\text{actual}}|]_{s,g,d,i} \\ \text{i.e. } \delta r &= [|r_{\text{desired}} - r_{\text{actual}}|]_{s} + [|r_{\text{desired}} - r_{\text{actual}}|]_{g} + \cdots \end{aligned}$$

(Please note here that δr is the differential of desired and actual value irrespective of positive or negative signs. This caution is essential to ensure positive and negative values do not balance out.)

And

$$\left[\sum r_{\max} \right]_{s,g,d,i} = 3 + 3 + 3 + 3 = 12.$$

So we have

$$V_{\text{aesthetic}} = 1 - \left(\left[\sum \delta r \right]_{s,g,d,i} \Big/ 12 \right). \tag{6}$$

This means that the aesthetic value of an artifact is zero when it is not able to generate any emotion at all and the sum of δr becomes equal to 12. Conversely, aesthetic value becomes 1 when the designer is able to produce the emotions exactly desired by market.

5.2 Summary

To summarize the chapter, we have established the coordinate system for emotions, identified the relevant emotions for the motorcycle design, located the emotions on the coordinate system and correlated the coordinates to the aesthetic value of a motorcycle.

This chapter forms the basis for the whole study in the book simply because this chapter establishes the fundamentals which define the discussions in the next chapters. To create an analogy, we cannot discuss ragas without the seven notes, we cannot discuss arithmetic without the digits and we cannot discuss literature without the 26 letters. In future chapters, we will be relating the elements of motorcycle design to this coordinate system.

References

1. Van Rompay, T.J.L., A.T.H. Pruyn, and P. Tieke. 2009. Symbolic meaning integration in design and its influence on product and brand evaluation. *International Journal of Design* 3 (2): 19–26.
2. Norman, Donald A. 2005. Emotional design, why we love (or hate) everyday things. *Basic Books, USA* 1: 21–24.
3. Krippendorff, Klaus. 2006. *The semantic turn, a new foundation for design*, 39–175. New York: Taylor and Francis.
4. Ross, Philip R., and A.G. Wensveen. 2010. Designing behaviour in interaction: Using aesthetic experience as a mechanism for design. *International Journal of Design* 4 (2): 3–13.
5. Chitturi, Ravindra. 2009. Emotions by design: A consumer perspective. *International Journal of Design* 3 (2): 7–17.
6. Plutchik. 1980. Nature of emotions (monograph on the internet). *American Scientist* 89(4). Available from www.fractal.org/Bewustzijns-Besturings-Model/Nature-of-emotions.html.
7. Vitters, Joar, Dyrdal, Guvnor Marie, and Roysamb Espen. 2005. Utilities and capabilities: A psychological account of two concepts and their relations to the idea of good life. (Conference proceedings) 2nd Workshop on Capabilities and Happiness, University of Milano, Italy: 1–23. Available from http://uit.no/getfile.php?pageId=1935&Field=38.
8. Reubenstrunk, Gerd. 1998. Emotional computers: computer model of emotions and their meaning for emotion-psychological research (Report on the internet). Available from www.ruebenstrunk.de/emeocomp/content.HTM.
9. Changeux, Jean-Pierre, Ricoeur, Paul, and DeBevoise, M.B. 2002 *What makes us think?*, 222–223. USA: Princeton University Press.
10. Barlinge, Surendra S. 2007. Bhartiya Soundarya Siddhant Ki Nayi Paribhasha. Bhartiya Gnanpith.
11. Levy, Pierre, Yamanaka, Toshimasa, and Tomico, Oscar. *Methods and means for Kansei design*. Available at http://pie.kansei.tsukuba.ac.jp/epierrot/publis/PLevy_ErgoDesign09.pdf.
12. Rauterberg, Matthias. 2006. How to assess the users' experience in cultural computing. Usability Professionals, 1–17.
13. Wollmer, Martin, Eyben, Florian, Douglas-Cowie, Ellen, and Cowie, Rodie. Abandoning Emotion Classes-towards Continuous Emotion Recognition with Modelling of Long-Rang Dependencies. Interspeech, ISST, September 22–26. Brisbane Australia, 597–600.
14. Dzogang, Fabon, Lisot, Marie-Jeanne, Rifqi, Maria, and Bouchon-Meunier, Bernadette.2010. Analysis of Texts' Emotional Content in a Multi-dimensional Space. In *International Conference on Kansei Engineering an Emotion Research*, 1–10.
15. Prinz, Jesse. 2007. Emotion and Aesthetic Value. In *Pacific APA Conference, San Francisco.*
16. Lee, SeungHee, Harada, Akira, and Stappers, Pieter Jan. Pleasure with Products: Design based on Kansei, 1–10. Available at http://www.keisen.com/documentos/pleasure%20with%20products.pdf.

17. Carter, Curtis. 1976. *Aesthetic values and human habitation: A philosophical and interdisciplinary approach to environmental aesthetics*, 81–95. American Values and Habitat: A Research Agenda.
18. Schummer, Joachim, MacLenan, Brace, and Taylor, Nigel. 2009. *Aesthetic values in technology and engineering design*. Handbook of the philosophy of science, Volume 9: Philosophy of technology and engineering sciences, 1031–1068.
19. Jones, G.R, Jones, Ilze, Gray, B.A., Parker, Bud, Coe, J.C., Burnham, J.B., and Geitner, N.M. 1975. A method for quantification of aesthetic values for environmental decision making. *Nuclear Technology* 25: 682–713.
20. Grimm, Brett. 2009. Quantifying visual effects of wind farms: A theoretical process in an evolving Australian landscape. Dissertation for Doctorate of Philosophy. School of Architecture, Landscape Architecture and Urban Design. University of Adelaide.

Chapter 6
Deconstructing Emotions into Design Factors

I am often accused of poking philosophy in those places where it is least required-automobile design being one of them. On the contrary, I firmly believe that one cannot understand engineering-specifically design—without understanding philosophy. The converse is also true, i.e., one cannot understand philosophy at its core without getting to the core of design which actually means understanding the interplay between the forces of nature. This is specifically true while we talk of design factors affecting the emotions. Because the relationship between the factors and the emotions is the relationship between the signifier and the signified, the core of deconstruction theory propounded by Derrida. If I put this complex theory in most simple words it is simply this- the sub-text of text is always the function of the text. While talking of design, this "text" is nothing but those design factors which manifest themselves physically and emotion is the sub-text which is derived by the human faculty of feelings. But here lies the rub. Most of us believe that human feeling is completely subjective and cannot be linked scientifically to design factors. Various idioms and phrases have been formed to illustrate this myth—the most famous being that beauty lies in the eyes of the beholder. For the sake of a scientific approach to design, this myth needs to be demolished—at least in the mind of the designer. So far as art and literature are concerned, this is easy. A melancholic song is supposed to generate sadness and it generates. A comedy is supposed to generate laughter and it generates. Now a melancholic song is constituted by specific elements of music like sequence and tone of notes. This means that a designed sequence and tone of notes generates a certain predetermined emotion. Similarly, in a comedy, elements like story-line and word-flow are designed to generate laughter. My simple point is—what is true for music and literature is true for design as well.

Let us again use the analogy of literature to understand the factors which dictate the emotions. Whether we take prose or poetry, there are fundamentally two things which decide emotion experienced by the reader (or viewer, if we consider cinema as a form of literature)—content and style. Finally, it is the interplay of these two factors which determines the ultimate emotion. Talking about automobiles—specifically motorcycles—"content" signifies all those engineering factors, which

S. Chandra, *Aesthetics: Quantification and Deconstruction*,
Design Science and Innovation, https://doi.org/10.1007/978-981-10-6235-3_6

create the context for the emotion in rider's mind. For example, when a rider looks at a bike with a large rake angle, large wheels and a V-twin engine with large cylinders, the context is generated in his mind as to how the bike will be behaving on the road and what kind of experience he will be getting while riding the bike. So we name this bunch of factors, which ultimately decide the behavior and performance of the bike, as "context." Beyond these contextual factors, we will be discussing the stylistic factors like color, form, unity, dynamism, graphics, etc., later in this chapter. But before we do all this, let us revisit the mathematics of emotions.

Referring to Fig. 5.3 and Table 5.4 in Chap. 5,once we have determined the emotions representing θ in the context of a motorcycle, i.e., glamor ($\theta = 0$), dynamism ($\theta = 90$), Intimidating ($\theta = 120$), and sober ($\theta = 270$), we face the task of determining the intensities of these emotions, i.e., "*r*." Since, the intensities for these emotions, i.e., s (sober), g (glamor), d (dynamism), and i (intimidation), is the weighted sum of different factors we can express it in terms of an equation:

$$[r]_{\mathrm{s,g,d,i}} = \left[\sum w_{\mathrm{x}} r_{\mathrm{x}}\right]_{\mathrm{s,g,d,i}} \tag{6.1}$$

where "*w*" is the weightage for various design factors, and "*x*" represents various design factors which we will establish in subsequent chapters.

Again, r_{x} for a design factor also is a weighted sum of different sub factors, the equation can be further broken down into:

$$[r_{\mathrm{x}}]_{\mathrm{s,g,d,i}} = \left[\sum w_{\mathrm{a}} r_{\mathrm{xa}}\right]_{\mathrm{s,g,d,i}} \tag{6.2}$$

where "*a*" represents various design factors which we will establish in this chapter.

6.1 Context

Zangwill [1] in his article "Music, Metaphor, and Emotion" contrasts the two theories of relationships between object and emotion, namely literalism (where emotion is direct literal product of the object) and metaphorism (where emotion is a derivative of object's metaphorical implications) and argues in favor of the later. Since a motorcycle is not an object directly constituted by words like a story or a poem or a song, this argument boils down to the fact that emotion generated by an object is the product of metaphors it evokes. Krippendorf [2] has classified metaphors as visual and verbal. Visual metaphor relates to the visual architecture of the object. In case of motorcycles, the examples for visual metaphors are step-through and step-over architecture, naked or full cowl body, etc. Verbal metaphors are more complex as they are the sub-text of the text. These words, which are directly not applicable to the object but are the sub-texts generated from the context. Let us take the word "muscular" for motorcycles. Clearly, motorcycles do not have muscles. But this word is evoked when a certain type of design is seen.

Fig. 6.1 Motorcycles with varying contexts

Let us look at the three motorcycle images above (Fig. 6.1). The first one is multi-cylinder cruiser with large thin wheels and a naked round head-light. Does this set of information generate an expectation about the experience; the rider will have on the road? Certainly. Exactly the same will be the answer for the two other bikes where the second one is multi-cylinder sports bike with big windscreen and shorter thicker wheels and the third one a simple single-cylinder with bare minimum parts and a small visor over a rectangular head-light. So, there are engineering informations which we grasp through our eyes and generate our expectations based on them. These expectations, then turn into emotions. For example, with a cruiser, you expect a powerful but quiet ride and feel the emotion of quiet heroism probably represented by Clint Eastwood. Similarly, the sport bike generates an expectation of speed and thrill leading to dynamism of Tom Cruise. The third one, with its simple commonsensical specification and bare essentials leads to the emotion of sobriety and simplicity probably represented by Dustin Hoffman. Fundamentally, the engineering information sets a context in the mind and the context generates the emotion. So, to be precise, in the context of a motorcycle, these verbal metaphors vary depending on the following factors:

(a) Visual architecture
(b) Type and construction of motorcycle, i.e., cruiser, commuter, or sports
(c) Front facia of the motorcycle constituting the head-light, speedometer, and visor
(d) Type and size of the wheel.

6.2 Color

To say that color is a something which affects emotion is stating the obvious. But it is too simple a proposition considering the complexity of color. What appears as one single color can have hundreds of variations. The complexity does not stop

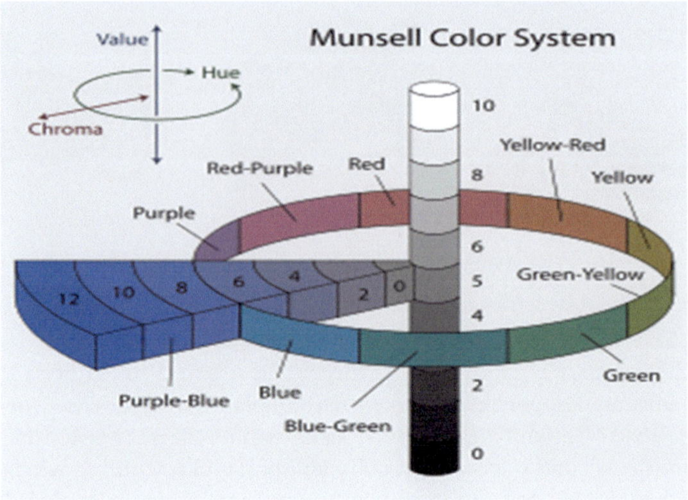

Fig. 6.2 Munsell's color wheel

here. Real life is never monochromatic. Even a combination of two colors makes the number of possible variations unmanageable. Here I am not getting into a poetic monolog about color. I am talking about the engineering of color—pure and simple.

The most complex issue about color is expressing it. To say that the color of a car body is blue is meaningless. Even light blue is too insufficient. There are hundreds of variation of light blue color with varying degrees of lightness and greyness. There have been extensive studies on how to express color. There are three theories which are mostly in vogue—Munsell's, Prang's and Ostwald's. All the three models express a given colors in terms of different scales. Most popular among them is Munsell's color wheel (Fig. 6.2), which expresses color in terms of three scales-hue, value, and chroma. Let us understand the three terms.

Hue: To put it simply, it is the name of the color like red, blue, green, etc. But as it happens everywhere, life is not as simple as it looks. In reality, very seldom we find pure or 100% red or 100% green. More often, it is a mix of colors. The best example is yellow color, which is nothing but a mix of red and green. Similarly, white is a mix of all colors and black is absence of all colors. Point is how to express it. Today, we have two systems prevalent systems—RGB (Red, green, blue) and CMYK (cyan, magenta, yellow, and black). The CMYK system is a relic of old printing age when they had to mix the inks to produce the desired color and the basic ink colors used to be cyan, magenta, yellow, and black. In the current digital age RGB system is used, which uses the primary colors. Here, the hue of a color is expressed by stating the presence of red, blue, and green on a scale of 0–255. Almost all hues in the world can be quantitatively expressed in terms of these three scales. For example, pure red is 255, 0, 0; pure green is 0, 255, 0 and pure blue is 0, 0, 255. Similarly, yellow, cyan, and magenta are 255, 255, 0; 0,255,255 and

255, 0, 255, respectively. White and black can be represented as 255, 255, 255, and 0, 0, 0. What you see in the Munsell's wheel is a simplified representation which is not exhaustive.

Of course, equipments in form of analyzers are available to exactly measure the quantity of R, G, and B on this scale.

Value: Every color has certain amount of whiteness inherently present in it. This amount of whiteness is expressed as value on a scale 0–10. If the value is 0, the complete absence of white makes it pitch dark or black. Similarly, the value of 10 makes it absolute white. This means any color other than black and white has value more than 0 and less than 10. Have a look.

Chroma: Chroma is also known as saturation and is a measure of its greyness and is expressed on a scale of 0–12. A chroma of 0 means absolute grey and a chroma of 12 means zero presence of grey. A higher chroma means more saturated color which in simple terms means depth and purity.

From manufacturing perspective, color is a product of either the coating process or finishing process (in case of uncoated parts). For plastics, it comes either from pigmenting (in-mold colors) or painting. For metallic surfaces, it comes either from plating or painting. In some cases, it is the natural color of the material itself.

But the visual impact of color is always coupled with the visual impact of the surface texture. We can define texture on two planes—technological and perceptual. Technologically, as the term "texture" is known in material science, it means a particular orientation distribution in microstructure. They go on to define the perceived texture as "the synthesis of physiological and psychological responses to the geometrical configuration (of microstructure) and the physical-chemical attributes of the surface". These two definitions are related as the perception is a result of geometrical configuration, but for our purpose, we will go with the definition of perceived texture only as the very purpose of this study is to know the perceptual impact.

In context of motorcycles, the textures of surfaces can be roughly classified as plain, rough, and glossy (It is a very rough classification, but if we go into finer classifications, it will need a separate study, outside the scope of this study. Moreover, with more classifications, the number of permutations and combinations along with colors will become too unwieldy for a survey). These texture result from two sources:

(a) **Coatings and finishing on metallic surfaces**: Chrome plating generated gloss. Zink plating generates glossy or plain textures. Paints generate glossy or plain textures. Powder coating generates all three kinds of surfaces. Moreover, the texture depends also on the surface finish of the surface.
(b) **Coatings and finishing on plastic surfaces**: Plastics are pigmented, painted or plated. In case of pigmenting, the texture depends on die-polishing, and grain-finish. In case of painting, it can be solid or glossy depending on the type of paint. Plating on plastics generates gloss.

The visual impact of texture on emotions can be studied only in combination with colors.

6.3 Form

Designer Reed [3] classifies the forms in two basic categories-geometric and organic which have a bearing on emotional aspect of an object. Mathematically, any surface consisting of curves defined by more than one equation is organic surface. But a motorcycle is not a single surface. It is formed by many objects, each of which is a combination of many surfaces. How these surfaces relate to each other is a major factor contributing to the emotion it generates. All the visible parts of a motorcycle can be divided in two categories, those constituting the style of a bike and the visible functional parts. Sure, the functional parts also play a major role in deciding the character of a motorcycle, but that aspect is covered in the factor "context." Here the styling parts, where form is designed to evoke an emotion are visor, fuel tank, side cover, and seat and the aspects of form which play a major role are space, fillets, and curvature.

6.4 Unity

Till now we have discussed factors which relate to individual parts but the emotion evoked by a motorcycle is not the sum of emotions evoked by all parts. There is something in the holistic scheme of parts which plays the deciding role. Unity, dynamism, and symmetry are such visual factors resulting from this holistic scheme. Whereas symmetry is inherent in the motorcycle design as the left and right side are almost always symmetric, unity and dynamism are complex concepts which need to be geometrically defined in the context of motorcycles. McClurg-Genevese [4] defines unity as a concept that describes the relationship between individual parts and the whole of a composition. Jirousek [5] defines it as an underlying principle that refers to the coherence of the whole, the sense that all of the parts are working together to achieve a common result. Lauer et al. [6] define unity of design as a concept wherein a congruity or agreement exists among the elements in a design; as though a visual connection beyond mere chance has caused them to come together. Zelanski et al. [7] define it as a property of art or design, where the whole will seem to be more than the sum of its parts

Unity in the context of motorcycle can be defined as the mutual compatibility between five factors—unity of form, continuity of parts, character lines, feature lines, and graphics:

(a) Unity of form: Two objects are having unity of form if they have either similar or complimentary forms. Form, as we know are either geometric or organic, can

Fig. 6.3 Example of unity of form

have one or more surface with positive–positive or positive–negative spaces. Besides, the similarity or complementation should be visibly distinct. The illustration (Fig. 6.3) is a good example of unity of form where all elements are having different sizes and colors but are having same form.

(b) Continuity: When two objects appear to be extensions of each other visually and coherently, they have continuity. The motorcycle illustration (Fig. 6.4) gives a good example of continuity, where fuel tank, seat, side cover, and rear cowl appear as extensions of each other.

(c) Feature lines: Feature lines are lines formed by visible boundaries of parts.

(d) Character lines: Character lines are distinctly visible lines formed by intersection between two surfaces.

(e) Graphics: Graphics, as we know, are visual predominantly two-dimensional images added to the visible surfaces.

Unity of design in a complex machine can be derived by combination of above parameters and unity between these parameters themselves.

Fig. 6.4 Continuity, character lines, and feature lines

6.5 Dynamism

One important factor needs to be noted. Motorcycle is a dynamic object and its dynamism cannot be established merely on the basis of static visual factors. The experience that a rider gets while driving the motorcycle plays an important role in evaluating the dynamism inherent in a vehicle. But, these aspects get covered in "context" factor, where the type of motorcycle covers the engine performance, acceleration, and the rider's posture.

Therefore, the scope of this factor "dynamism" is limited only to its visual aspect. Visually dynamism can be defined as the element of visual tension created by the impression that the visual we are seeing is not going to remain static and something is going to change immediately. This tension is generated by three elements of the visual:

(a) **The divergence and alignment of lines**: We will call it visual dynamism. We can illustrate this example by contrasting the image of a standing tiger with a galloping tiger. Whereas in the first case the four legs of the tiger as vertically parallel and the torso is horizontal, in the case of the galloping tiger the four legs are diverging and the torso is inclined upwards in the direction of the movement. This divergence immediately makes us feel the possibility of imminent change. Here also character lines, feature lines, and graphics play an important role. But, what matters more here is their orientation and alignment with each other.

(b) **The visual impression of center of gravity**: This impression is generated by the height and length ratio. The more is this ratio, the more is the perceived height of the center of gravity and thus the possibility of falling.

(c) **The visual imbalance**: Any balanced visual gives us the sense of stability. Conversely, visual imbalance creates the sense of instability and hence generates the visual tension.

Now, these two factors unity and dynamism have a complex relationship. In fact, the process of generation of emotion results from the interplay of these two factors. We will discuss about this interplay in detail in later chapters when the actual impact of the factors on emotion is discussed.

6.6 Graphics

Though, graphics are not an engineering parameter in the strict definitive sense, it is an important contributor to the overall aesthetics of a motorcycle. Sometime it has been observed that the same motorcycle with different graphics revolutionizes its aesthetic character. Motorcycle graphics have two important elements—logo and illustrations. Sometimes, they are separated and sometimes integrated.

Logo: Logos are used to inform the user about the manufacturing company and the product name. But more than this purpose, they are important carriers of company and product's brand image. They are either three-dimensional objects (either plastic or metallic) or wrapped around a prominently visible parts.

Typography: The most important aspect of logo is its typography (both for model name and company name). Typography is a science developed through centuries and various typographical styles have evolved. The different stylistic aspects of typography, which decide the character of the bike, are:

(a) The size
(b) The continuity or flow
(c) The sharpness of joints
(d) The number of geometric elements

   a. The number of straight elements
   b. The number of curved elements

All these elements with their various combinations are covered in typographical styles evolved through years. Sarkar [8] describes these styles to be divided into four groups (Table 6.1).

What impacts the visual aesthetics and determines the emotional character of the motorcycle is various combinations of logo and illustrations, their visual design concept and the color combinations.

Here, we have taken only a dip-dive into the factors to have a fundamental understanding about them. The real challenge is to quantify them, find a method to measure or calculate the quantified values and study their impacts on emotion. We will discuss all of these in further chapters.

Table 6.1 Hierarchy of typographical styles

Classical	Text	
	Roman	Old style
		Transitional
		Modern
Lineal	Sans Serif	
	Square Serif	
Fancy	Ornamental	
	Novelty	
Handwritten	Cursive	
	Script	

References

1. Zangwill, Nick. 2007. Music, metaphor and emotion. *The Journal of Aesthetics and Art Criticism* 65(4): 391, 394 (Fall).
2. Krippendorff, Klaus. 2006. *The Semantic Turn, a New Foundation for Design*, 39–175. New York: Taylor and Francis.
3. Hannah, Gail Greet. 2002. *Elements of Design, Rowena Reed Kostfellow Foundation*, 9–118. USA: Princeton Architectural Press.
4. McClurg-Genevese, and Joshua David. 2005. *The Principles of Design*. Available at http://www.digital-web.com/articles/principles_of_design.
5. Jirousek, Charlotte. 1995. *Art Design and Visual Thinking*. Available at: http://char.txa.cornel.edu.
6. Lauer, David A., and Stephen Pentak. 2007. *Design Basics*, 28–226. USA: Thomson Wadsworth.
7. Zelanski, Paul., and Fisher, Mary Pat. 1996. *Design Principles and Problems*, 34–35. USA: Wadsworth Cengage Learning
8. Sarkar, N.N. 1995. *Art and Production. Chapter 4 &5*, 34–65. Delhi: Sagar Publications.

Chapter 7
The Methodology

Once we have understood how to express aesthetics in quantities, the next question is how to estimate the quantities. Talking about principles and fundamental concepts is fine but it is like talking in thin air. It is like knowing that universe is ruled by gravity but not knowing how to calculate the gravitational force. More importantly and interestingly, most often we face the problem in reverse where we know the answer and we have to find the question. I mean, we know the emotion mix but the designer has to find out the design attributes which will take him there. Now, it is not that we are the first people on earth to meditate on this. Nagamachi [1] has done an extensive work on this and this study is called Kansei engineering. Let us understand Kansei engineering first.

7.1 The Approach

Kansei Engineering System basically is a methodology to develop the engineering attributes from the emotional requirements of an artifact. The following diagram (Fig. 7.1) explains the fundamental basis of this approach.

What we get from this approach is the desired engineering properties if we want to generate a specific psychological feeling or emotion. In other words, if we want our motorcycle to develop a certain degree of glamor combined with a certain degree of intimidation, this approach should lead us to determine what kind of space combination, body construction, color scheme, or fillet radii should be designed (see Fig. 7.2). Is this not precisely what we need?

Schutte [2], a Swedish researcher has done further work on Kansei engineering and suggested following different approaches to it:

(a) Approach 1: Category classification (Fig. 7.3)
(b) Approach 2: Kansei Engineering System (Fig. 7.4)
(c) Approach 3: Hybrid KES (Fig. 7.5).

S. Chandra, *Aesthetics: Quantification and Deconstruction*,
Design Science and Innovation, https://doi.org/10.1007/978-981-10-6235-3_7

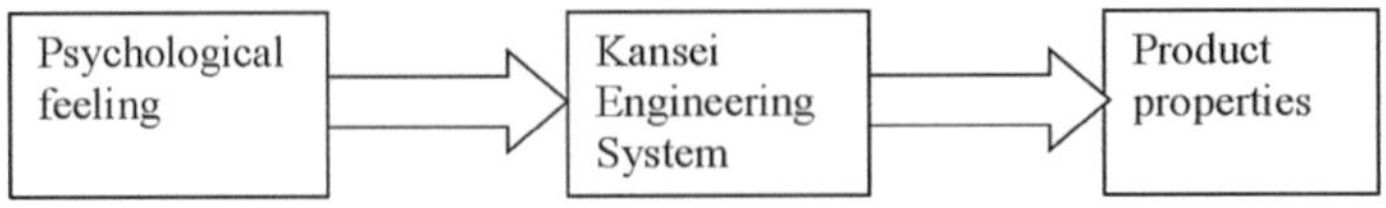

Fig. 7.1 Kansei engineering framework

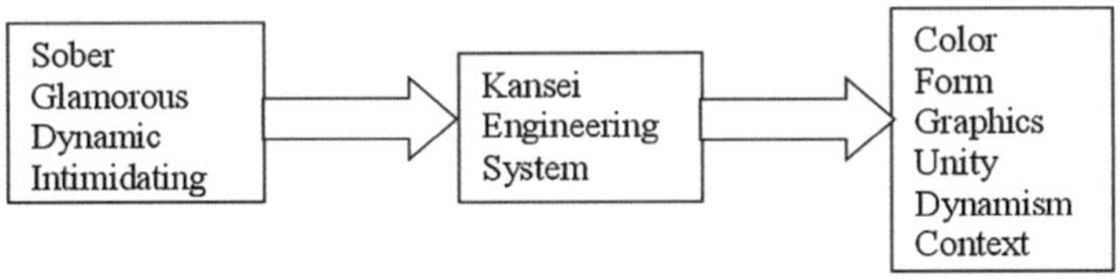

Fig. 7.2 Kansei engineering framework for bike design

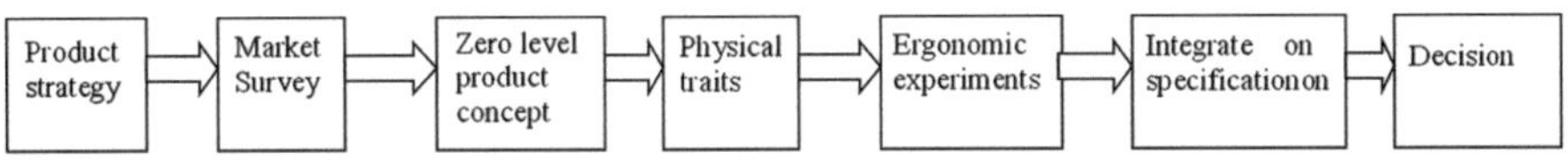

Fig. 7.3 Kansei engineering, approach 1

Fig. 7.4 Kansei engineering, approach 2

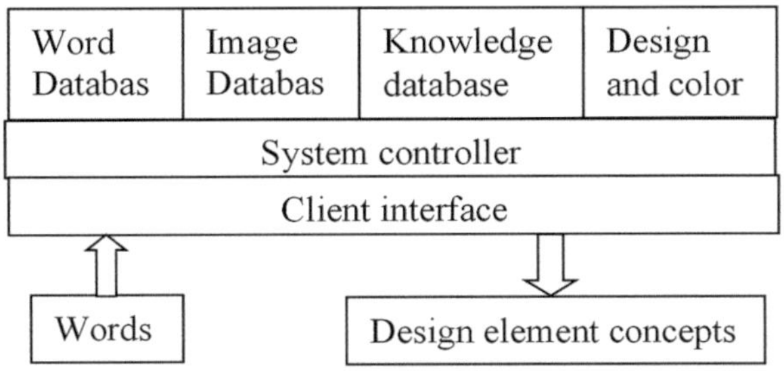

Fig. 7.5 Kansei engineering approach 3

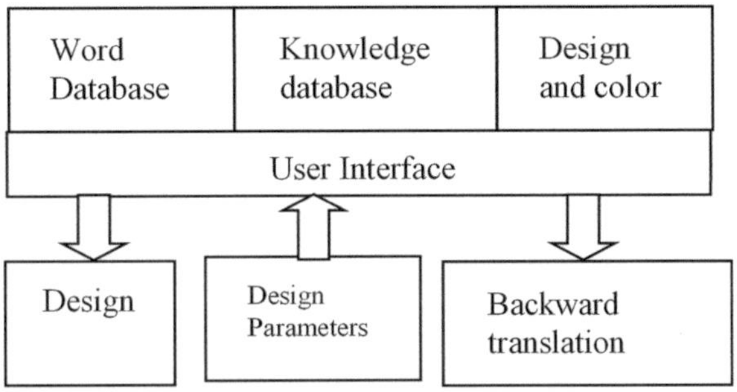

Based on the inputs we have and the output we desire, it is the third approach which appears to fulfill what we need because here the databases give us an indication about how to design and a backward translation about the way the design is able to achieve its stylistic and emotion goals. Let us develop the idea.

Now, based on the third approach, the fundamental objective of the study gets clearly sketched out. The user interface has to develop the relationship between the emotions and the design attributes based on the word and design databases. This requires the collection, collation, and analysis of data on generation of emotions as a result of response to design inputs. To do that, let us revisit Eqs. (2) and (3) in Chap. 5.

$$[r]_{\mathrm{s,g,d,i}} = \left[\sum w_x r_x\right]_{\mathrm{s,g,d,I}} \quad (1)$$

where r and w are the intensities of the four emotions and weightage for a design factor respectively.

Now, we can exactly describe "x" as l (color), f (form), t (context), g (graphics) and u (unity and dynamism).

We can further describe the sub-factors embedded in Eq. (1), i.e.,

$$[r_x]_{\mathrm{s,g,d,i}} = \left[\sum w_{\mathrm{a}} r_{\mathrm{xa}}\right]_{\mathrm{s,g,d,i}}$$

as

$$[r_l]_{\mathrm{s,g,d,i}} = \left[\sum w_{\mathrm{la}} r_{\mathrm{la}}\right]_{\mathrm{s,g,d,i}} \quad (2)$$

where la = color and texture as sub-factors of color.

$$[r_f]_{\mathrm{s,g,d,i}} = \left[\sum w_{\mathrm{fa}} r_{\mathrm{fa}}\right]_{\mathrm{s,g,d,i}} \quad (3)$$

where fa = fuel tank, seat, side cover and visor as sub-factor of form.

$$[r_t]_{\mathrm{s,g,d,i}} = \left[\sum w_{\mathrm{ta}} r_{\mathrm{ta}}\right]_{\mathrm{s,g,d,i}} \quad (4)$$

where ta = type, front facia and wheels as sub-factors of context.

$$[r_g]_{\mathrm{s,g,d,i}} = \left[\sum w_{\mathrm{ga}} r_{\mathrm{ga}}\right]_{\mathrm{s,g,d,i}} \quad (5)$$

where ga = typography and stripes as sub-factors of graphics.

Since unity and dynamism has no sub-factor, Eq. (1) holds good.

To understand and visualize these equations let us put it in a tabular format (Table 7.1) which makes the calculation pretty obvious. You can see that the

Table 7.1 Calculating emotion scores

	Weightage (*x*)	*S*	*G*	*D*	*I*	
(a) Color						
Color	W_{la}	$[r_{la}]_s$	$[r_{la}]_g$	$[r_{la}]_d$	$[r_{la}]_i$	
Texture	W_{lb}	$[r_{lb}]_s$	$[r_{lb}]_g$	$[r_{lb}]_d$	$[r_{lb}]_i$	
Weighted Aggregate = $r_{l(s,g,d,i)}$		$\sum W_{ln}[r_{ln}]_s$	$\sum W_{ln}[r_{lx}]_g$	$\sum W_{ln}[r_l]_d$	$\sum W_{ln}[r_l]_i$	n = a, b
(b) Form						
	Weightage (x)	*S*	*G*	*D*	*I*	
Fuel tank	W_{fa}	$[r_{fa}]_s$	$[r_{fa}]_g$	$[r_{fa}]_d$	$[r_{fa}]_i$	
Seat	W_{fb}	$[r_{fb}]_s$	$[r_{fb}]_g$	$[r_{fb}]_d$	$[r_{fb}]_i$	
Visor	W_{fc}	$[r_{fc}]_s$	$[r_{fc}]_g$	$[r_{fc}]_d$	$[r_{fc}]_i$	
Side cover	W_{fd}	$[r_{fd}]_s$	$[r_{fd}]_g$	$[r_{fd}]_d$	$[r_{fd}]_i$	
Weighted aggregate = $r_{f(s,g,d,i)}$		$\sum W_{fn}[r_{fx}]_s$	$\sum W_{fn}[r_f]_g$	$\sum W_{fn}[r_{fn}]_d$	$\sum W_{fn}[r_{fn}]_i$	n = a, b, c, d
(c) Graphics						
	Weightage (*x*)	*S*	*G*	*D*	*I*	
Stripes	W_{ga}	$[r_{ga}]_s$	$[r_{ga}]_g$	$[r_{ga}]_d$	$[r_{ga}]_i$	
Typography	W_{gb}	$[r_{gb}]_s$	$[r_{gb}]_g$	$[r_{gb}]_d$	$[r_{gb}]_i$	
Weighted aggregate = $r_{g(s,g,d,i)}$		$\sum W_{gn}[r_{gn}]_s$	$\sum W_{gn}[r_{gn}]_g$	$\sum W_{gn}[r_{gn}]_d$	$\sum W_{gn}[r_{gn}]_i$	n = a, b
(d) Context						
	Weightage (*x*)	*S*	*G*	*D*	*I*	
Type	W_{ta}	$[r_{ta}]_s$	$[r_{fa}]_g$	$[r_{fa}]_d$	$[r_{fa}]_i$	
Fr facia	W_{tb}	$[r_{tb}]_s$	$[r_{tb}]_g$	$[r_{tb}]_d$	$[r_{tb}]_i$	
Tires and wheels	W_{tc}	$[r_{tc}]_s$	$[r_{tc}]_g$	$[r_{tc}]_d$	$[r_{tc}]$i	
Weighted aggregate = $r_{t(s,g,d,i)}$		$\sum W_{tn}[r_{tn}]_s$	$\sum W_{tn}[r_{tn}]_g$	$\sum W_{tn}[r_{tn}]_d$	$\sum W_{tn}[r_{tn}]_i$	n = a, b, c

(continued)

Table 7.1 (continued)

	Weightage (x)	S	G	D	I	
(*e*) *Unity and dynamism*						
		S	G	D	I	
Score = $r_{u(s,g,d,i)}$		r_{us}	r_{ug}	r_{ud}	r_{ui}	
			r			
	Weightage (x)	S	G	D	I	
Color	W_l	r_{ls}	r_{lg}	r_{ld}	r_{li}	
Form	W_f	r_{fs}	r_{fg}	r_{fd}	r_{fi}	
Graphics	W_g	r_{gs}	r_{gg}	r_{gd}	r_{gi}	
Context	W_t	r_{ts}	r_{tg}	r_{td}	r_{ti}	
Unity and dynamism	W_u	r_{us}	r_{ug}	r_{ud}	r_{ui}	
Final weighted aggregate	$r_{(s,g,d,i)}$	$\sum W_x r_{xs}$	$\sum W_x r_{xg}$	$\sum W_x r_{xd}$	$\sum W_x r_{xi}$	x = l, f, g, t, u

emotion score for each design sub-factor (e.g., $[r_{la}]_s$) is aggregated into an emotion score for the design factor (r_{ls}) after taking into account the weightages (W_{la}) for the sub-factors and these score are aggregated into final scores ($r_{s,g,d,i}$) after taking into account the weightages for the design factors (W_l, W_f, etc.).

Now, this table gives us an idea of the word and design databases to be generated and analyzed. We can see that fundamentally two kinds of design databases have to be generated to establish—(a) the weightages for sub-factors and factors and (b) the emotion scores i.e. $r_{l(s,g,d,i)}$, etc. The database for weightage is to be created by creating images with differences in design factors as if the other factors do not change. For example if weightages for factor X (which has values X_a, X_b, and X_c) and factor Y (which has values Y_a, Y_b, and Y_c) has to be established, we have to create a matrix of images with three rows with X_a, X_b, and X_c and three columns with Y_a, Y_b, and Y_c. The following diagram illustrates this. In this diagram, images have been created for deciding the weightages for sub-factors of context. These sub-factors are type of motorcycle (standard or sports), visor (with visor or without visor) and tires and wheels (narrow and wide, cast and spoke). So the image matrix covers all these variations as shown in Table 7.2.

Now, these images are simply to be shown to the respondents and the respondents have simply this to tell—which is the emotion that comes to your mind? Is it sobriety, glamor, dynamism, or intimidation? You may ask—why only one emotion when actually it is a mix of emotions. Answer is simple—simplicity. The process has to be simple which does not leave the respondent bogged down. Now once, the respondents have identified the emotion, they have to quantify the intensity of the emotion on a Likert scale of 1–3 (again, it cannot be zero because this is the emotion which has been identified).

Similar process is repeated for emotion score where design data for each sub-factor is generated and presented to the respondents who have to respond in this same fashion, that we have just discussed. We will discuss more on this in the respective chapters on each design factor.

Table 7.2 Image matrix for weightage

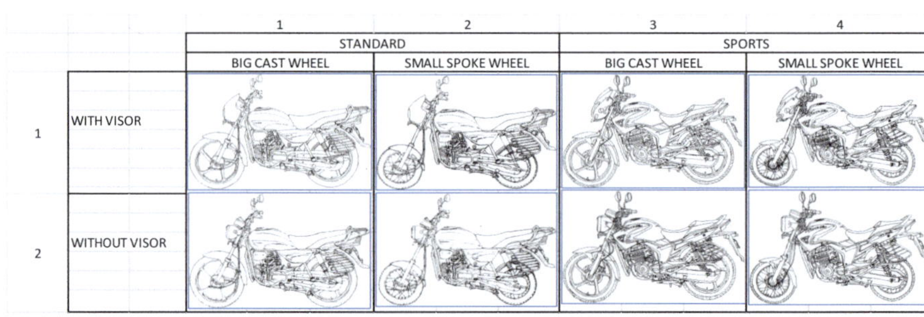

7.2 Mathematical Modeling

Weightage: Having collected the *S, G, D,* and *I* emotion scores for each of the image, we need to calculate the weightages for the factors or sub-factors (as applicable for the exercise). For this we need simply to calculate the impact of variation in each factor (X_1, X_2, etc.) by aggregating separately the difference in *S, G, D,* and *I* scores.

Example: Let us take the example as shown in Table 7.2. Here we need to find the weightages for context sub-factors—visor (with visor/without visor), wheel (cast/spoke) and category (standard/sports). The scores for *S, G, D,* and *I* for cast and spoke wheels are a_{11cs}/a_{21ss}, a_{11cg}/a_{21sg} … (The nomenclature for the images is column number followed by row number in subscript to letter "*a*" (example-a_{11}).

Impact of wheel variation $X1 = (a_{11cs}–a_{21ss}) + (a_{11cg}–a_{21sg}) + \cdots + (a_{12cs}–a_{22ss}) + (a_{12cg}–a_{22sg}) + \ldots$

Similarly we can get impact of visor variation (X_2) and category variation (X_3). Now the weightage for wheel is calculated as $X_1/(X_1 + X_2 + X_3 + X_4 + X_5 + X_6)$. Similarly weightages for other factors can be calculated.

Here, it is important to understand that this book is more about methodology than about outcomes though for most parts in the book, we have discussed outcomes. But the point is to realize that outcomes can change based on many factors like culture and engineering evolutions. But, the principles of aesthetics will not change and therefore the methodology will not change. This method to establish the weightages can be repeated in any age and in any other society with different outcomes. The reason, I am emphasizing this point is that the outcomes that I am discussing in this book are applicable in the particular context only. Now, the weightages established in my study are given in Table 7.3.

Since the impact of unity and dynamism has been combined (we will discuss the logic for this in Chap. 11), the weightage table has been modified as (Table 7.4).

Similarly the weightage for color and texture (Table 7.5), form (Table 7.6), Context (Table 7.7) and typography and graphics (Table 7.8) have been calculated as:

Impact scores: For evaluating the overall *S, G, D* and *I* score, the table shown in Table 7.1 can be upended based on the weightages determined (Table 7.9). Table 7.1 shows cells without numerical values for *S, G, D,* and *I*, which are to be filled while calculating the overall scores for a motorcycle. The values to be filled are to be generated after the surveys (The values shown in the following table are for example only).

Table 7.3 Weightages for design elements (unaltered)

Factor	Weightage %
Color	12
Form	18
Graphics	15
Context	25
Unity	16
Dynamism	14

Table 7.4 Weightages for design elements (alltered)

Factor	Weightage %
Color	12
Form	18
Graphics	15
Context	25
Unity and dynamism	30

Table 7.5 Weightages for color and texture

Color	50%
Texture	50%

Table 7.6 Weightages for form of components

Fuel tank	31%
Seat	34%
Visor	31%
Side cover	4%

Table 7.7 Weightages for context elements

Type	54%
Fr facia	23%
Tires and wheels	23%

Table 7.8 Weightages for typography and graphics

Typography	75%
Graphics	25%

Table 7.9 Motorcycle evaluation table

	Weightage	*S*	*G*	*D*	*I*	Details
(a) Color						
Body color	0.33	1.82	1.36	1.09	0.09	
Body/seat comb	0.33	0.62	1.45	1.45	0.93	
Body/stripe comb	0.33	0.62	1.45	1.45	0.93	
Mean(a)	0.5	1.02	1.42	1.33	0.65	
Texture(b)	0.5	0.98	2.43	2.00	2.15	
Weighted mean		0.998677	1.923033	1.664922	1.397682	
(b) Form						
	Weightage	*S*	*G*	*D*	*I*	Details
Fuel tank	0.31	1.31	0.95	0.58	0.21	
Seat	0.34	1.12	1.04	0.87	0.04	
Visor	0.31	0.47	1.66	1.28	1.09	
Side cover	0.04	0.59	0.83	2.26	0.83	

(continued)

Table 7.9 (continued)

	Weightage	*S*	*G*	*D*	*I*	Details
Weighted mean		0.9562	1.1959	0.9628	0.8162	
(*c*) *Graphics*						
	Weightage	*S*	*G*	*D*	*I*	
Typography	0.75	0.855	0.57	0.52	0.585	N14/H21
Graphics	0.25	0.92	0.68	1.99	0.44	
Weighted mean		0.87125	0.5975	0.8875	0.54875	
(*d*) *Context*						
	Weightage	*S*	*G*	*D*	*I*	
Type	0.54	0.14	0.74	2.87	1.39	
Fr facia	0.23	0.27	1.86	1.25	0.3	
Tires and wheels	0.23	0.2	1.9	0.93	0.24	
Weighted mean		0.1837	1.2644	2.9973	0.8748	
(*e*) *Unity and dynamism*		*S*	*G*	*D*	*I*	
		0.64	1.36	2.25	1.88	U3D3
(*f*) *Overall*				*r*		
	Weightage (*x*)	*S*	*G*	*D*	*I*	
Color	0.12	0.998677	1.923033	1.664922	1.397682	
Form	0.18	0.9562	1.1959	0.9628	0.8162	
Graphics	0.15	0.87125	0.5975	0.8875	0.54875	
Context	0.25	0.1837	1.2644	2.9973	0.8748	
Unity and dynamism	0.3	0.64	1.36	2.25	1.88	
Weighted mean		0.66057	1.259751	1.930545	1.17965	

For calculating the individual *S*, *G*, *D*, and *I* scores, for a particular survey sheet aggregate and tabulate all scores for *S*, *G*, *D*, and *I* and divide each of the aggregate score by the number of respondents and tabulate for all sheets. Finally, after equating maximum of the scores to 3.00, adjust each of the individual scores accordingly by multiplying each score by 3/max score.

Let us consider an example where there are five images relating to a specific design factor and five people have responded to the survey. After calculating the mean scores, we find that the score vary from 0.6 to 2.8. Now we are faced with a complication here. Suppose for a factor this score varies from 0.3 to 1.4. For another design factor this range is from 1.2 to 2.9 and for yet another factor this ranges from 0.1 to 1.9. This creates an imbalance as a design with maximum impact of factor A cannot be equated with maximum impact of factor B or C. This means the ranges need to be equalized where maximum score for each factor stands at the same level. This is done by calculating the weighted score as shown in the example (Table 7.10). One actual calculation sample has been given in appendix.

Table 7.10 SGDI score calculation method

No of respondents: (n = 5 e.g.)												
Sheet no	Aggregate score (a)				Mean score ($x = a/n$)				Weighted score ($X = x*3/x_{max}$)			
	S	G	D	I	S	G	D	I	S	G	D	I
1	12	10	7	4	2.4	2	1.4	0.8	2.57	2.14	1.50	0.86
2	5	8	11	7	1	1.6	2.2	1.4	1.07	1.71	2.36	1.50
3	14	11	3	10	2.8	2.2	0.6	2	3.00	2.36	0.64	2.14
4	7	12	7	13	1.4	2.4	1.4	2.6	1.50	2.57	1.50	2.79
5	3	4	12	14	0.6	0.8	2.4	2.8	0.64	0.86	2.57	3.00

7.3 Verification and Loop Closure

As I have stated in earlier pages, this book is about methodology and not outcomes. And the process of establishing methodology is not completed without the process of loop closure. This means, the results arrived by the methodology have to be somehow verified against actual results. So fundamentally it amounts to—calculating the results through the established methodology, collecting actual results, comparing them and tweaking the methodology for the two to match. This can be explained graphically through the following diagram (see Fig. 7.6). Let us understand in detail.

Step 1 **Identification of motorcycles**: This needs to be done very carefully. You need to identify the market, the culture, the geography and the technological scope for which you have established the methodology and pick the bikes representing the whole width of that scope. Moreover

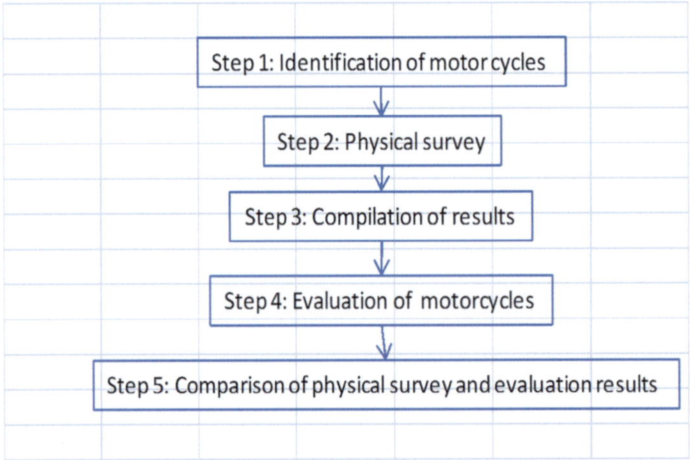

Fig. 7.6 Verification methodology

you have to guard against the evolutionary mismatch meaning that bikes should not belong to different eras.

Step 2 **Physical survey**: This most important step involves following steps:

(a) Physically lining up all motorcycles at one location.
(b) Preparation of survey sheet, listing all motorcycles and options available to participants.
(c) To collect all respondents and to brief them about the objectives of the survey and instructions.
(d) Respondents filling the survey sheet and giving their overall rating about the motorcycle. Here they have to rate the motorcycles as one of the characters and give the intensity of the character.

Step 3 **Compilation of results**: Respondents responses are to be tabulated and calculated as per the following Table 7.11.

Step 4 **Evaluation of motorcycles**: Now, evaluate each motorcycle using the evaluation table and the SGDI scores of each motorcycle is to be calculated.

Step 5 **Comparison of physical survey and evaluation results**: Now, tabulate the results of step 3 and step 4 against each other and compare.

This comparison will give many insights. Most important of the insights it can give is the efficacy of the very method we have evolved. Then, it will give an opportunity to adjust the weightages. Finally, the comparison will become an important tool to understand the market. One belief I have always held is that though this book is meant for motorcycles, this methodology can be applied to any artifact. For this very reason, the comparative study will prove to be a very effective tool for studying the market for any artifact.

Table 7.11 Physical survey result compilation

No of respondents: (n = 5 e.g.)												
Sheet no	Aggregate score (a)				Mean score ($x = a/n$)				Weighted score ($X = x*3/x_{max}$)			
	S	G	D	I	S	G	D	I	S	G	D	I
1	12	10	7	4	2.4	2	1.4	0.8	2.57	2.14	1.50	0.86
2	5	8	11	7	1	1.6	2.2	1.4	1.07	1.71	2.36	1.50
3	13	11	3	10	2.6	2.2	0.6	2	2.79	2.36	0.64	2.14
4	7	12	7	13	1.4	2.4	1.4	2.6	1.50	2.57	1.50	2.79
5	3	4	12	14	0.6	0.8	2.4	2.8	0.64	0.86	2.57	3.00
6	12	14	5	9	2.4	2.8	1	1.8	2.57	3.00	1.07	1.93

Appendix 1: Weightage Calculation for Context Sub-factors

		AVERAGE
11	S	1.52
	G	0.27
	D	0.02
	I	0.02
12	S	0.98
	G	0.40
	D	0.15
	I	0.02
13	S	0.30
	G	0.65
	D	0.84
	I	0.36
14	S	0.18
	G	1.00
	D	0.41
	I	0.25
21	S	1.30
	G	0.40
	D	0.24
	I	0.12
22	S	1.15
	G	0.18
	D	0.20
	I	0.10
23	S	0.41
	G	1.03
	D	0.32
	I	0.25
24	S	0.48
	G	0.44
	D	0.74
	I	0.21

Visor

Sheet		a	Sheet		b	a-b	abs
11	S	1.52	21	S	1.30	0.22	0.22
	G	0.27		G	0.40	-0.12	0.12
	D	0.02		D	0.24	-0.22	0.22
	I	0.02		I	0.12	-0.09	0.09
12	S	0.98	22	S	1.15	-0.17	0.17
	G	0.40		G	0.18	0.22	0.22
	D	0.15		D	0.20	-0.05	0.05
	I	0.02		I	0.10	-0.08	0.08
13	S	0.30	23	S	0.41	-0.11	0.11
	G	0.65		G	1.03	-0.39	0.39
	D	0.84		D	0.32	0.53	0.53
	I	0.36		I	0.25	0.11	0.11
14	S	0.18	24	S	0.48	-0.31	0.31
	G	1.00		G	0.44	0.56	0.56
	D	0.41		D	0.74	-0.33	0.33
	I	0.25		I	0.21	0.04	0.04
							3.55

Motorcycle type

Sheet		a	Sheet		b	a-b	abs
11	S	1.52	13	S	0.30	1.22	1.22
	G	0.27		G	0.65	-0.38	0.38
	D	0.02		D	0.84	-0.82	0.82
	I	0.02		I	0.36	-0.34	0.34
12	S	0.98	14	S	0.18	0.80	0.80
	G	0.40		G	1.00	-0.60	0.60
	D	0.15		D	0.41	-0.26	0.26
	I	0.02		I	0.25	-0.23	0.23
21	S	1.30	23	S	0.41	0.88	0.88
	G	0.40		G	1.03	-0.64	0.64
	D	0.24		D	0.32	-0.07	0.07
	I	0.12		I	0.25	-0.13	0.13
22	S	1.15	24	S	0.48	0.66	0.66
	G	0.18		G	0.44	-0.26	0.26
	D	0.20		D	0.74	-0.54	0.54
	I	0.10		I	0.21	-0.11	0.11
							7.94

Wheels

Sheet		a	Sheet		b	a-b	abs
11	S	1.52	12	S	0.98	0.54	0.54
	G	0.27		G	0.40	-0.13	0.13
	D	0.02		D	0.15	-0.13	0.13
	I	0.02		I	0.02	0.00	0.00
13	S	0.30	14	S	0.18	0.13	0.13
	G	0.65		G	1.00	-0.35	0.35
	D	0.84		D	0.41	0.43	0.43
	I	0.36		I	0.25	0.11	0.11
21	S	1.30	22	S	1.15	0.15	0.15
	G	0.40		G	0.18	0.21	0.21
	D	0.24		D	0.20	0.04	0.04
	I	0.12		I	0.10	0.01	0.01
23	S	0.41	24	S	0.48	-0.07	0.07
	G	1.03		G	0.44	0.60	0.60
	D	0.32		D	0.74	-0.42	0.42
	I	0.25		I	0.21	0.04	0.04
							3.36

∑∑ differential= ∑δVisor +∑δ Type+∑δ Wheels = 4.36+0.49+4.73= 14.85

$\text{Weightage}_{\text{visor}}$

.=3.55/14.85= 23.9 %

$\text{Weightage}_{\text{type}}$

.=7.94/14.85=53.5 %

$\text{Weightage}_{\text{wheels}}$

.=3.36/14.85= 22.6 %

Appendix 2: SGDI Calculation for Context Sub-factor Front Facia

	S		*G*		*D*		*I*	
	Survey score	Weighted score	Survey score	Weighted score	Survey score	Weighted score	Survey score	Weighted score
Sheet	*x*	*X*	*x*	*X*	*x*	*X*	*x*	*X*
1	1.72	3.00	0.00	0.00	0.04	0.08	0.07	0.11
2	1.52	2.65	0.22	0.38	0.15	0.27	0.07	0.11
3	1.04	1.82	0.37	0.64	0.15	0.27	0.07	0.11
4	1.41	2.46	0.13	0.23	0.07	0.11	0.04	0.08
5	1.30	2.28	0.72	1.25	0.15	0.27	0.00	0.00
6	0.98	1.71	0.83	1.44	0.17	0.30	0.00	0.00
7	0.35	0.61	0.63	1.10	0.67	1.18	0.11	0.19
8	0.02	0.04	1.11	1.93	0.65	1.14	0.07	0.11
9	0.15	0.27	1.07	1.86	0.72	1.25	0.17	0.30

References

1. Nagamachi, Mitsuo. 2002. Kansei enginnering as a powerful consumer-oriented technology for product development. *Applied Ergonomics* 33: 289–294.
2. Schutte, Simon. 2002. Designing Feelings into Products. Linkoping Studies in Science and Technology. Thesis No 946.

Chapter 8
The Context of Emotions

Tigers are beautiful. Snakes are also beautiful. Both have beautiful color-schemes, beautiful patterns, and movements absolutely stunning. Still both of them generate fear instead of glamor. The reason is obvious. It is the memories and narrative that both of them generate that is fearful. And these memories and narrative generate, what Krippendorf calls meaning of artifact in language. So, meaning of artifacts in language concerns how artifacts come to live in the narratives of the stakeholders. The theory that drives the meaning of artifacts is essentially a social theory, because language directs the concerns of at least two people but more typically of a community. This theory addresses the second order understanding that individuals have of each other and of the artifact. Following aspects constitute the essential features of meaning in language.

Category: This means how human beings categorize the artifacts as sub-category, category, and super-category. So far as motorcycles are concerned, it started as part of super-category of transport. When it started, automobile itself was a category which had to compete with other modes of transport like bicycles, horse carriages, etc. Within this, two wheelers became a sub-category. As times changed, automobiles became the de facto mode of transport and two wheelers became a category within the super-category of automobiles and motorcycles became a sub-category. Further down the line, motorcycles also had the sub-categories of cruiser, sports, and standard. Language being a dynamic entity, these shifting sands of time can be graphically seen in Fig. 8.1.

Characters: Each artifact assumes a character in language. This character evolves through the social perceptions and collective memories of a society through its history. Companies strive to develop products that carry distinctive references to the character of their brands (Karjalainen). This idea that objects have characters are all language attributions, acts performed in our language and reflect perceptual, emotional, and experiential coordinations. Motorcycle characters like robust, cool, and glamorous are all relative and creations of our perceptions. There is a difference between physical properties and characters in that physical property like weight is a constant for an object whereas character like heaviness depends on the context in which it is being referred to. The super fuel efficient bikes of 1910 will be

S. Chandra, *Aesthetics: Quantification and Deconstruction*,
Design Science and Innovation, https://doi.org/10.1007/978-981-10-6235-3_8

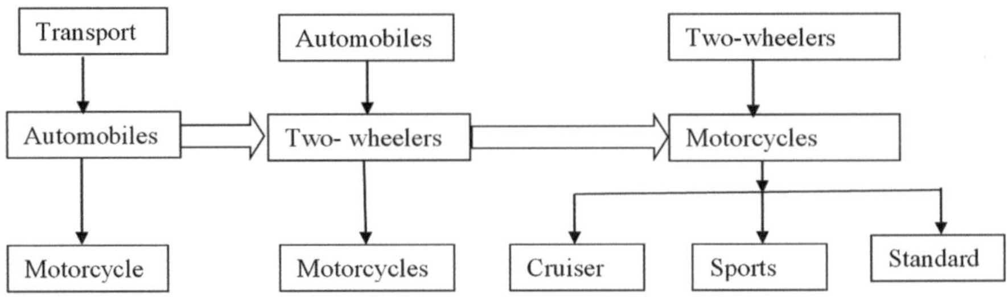

Fig. 8.1 Evolution of motorcycle categories

considered fuel guzzlers today, whereas the engineering parameter BSFC (Brake Specific Fuel Consumption) remains the same. Characters have following characteristics:

1. They are different from properties in that they are creations of human perceptions.
2. They are relative and comparable.
3. They exist in opposites and continuums like hard-soft, heavy-light, robust-fragile, etc.

Now, in context of motorcycle, we need to establish the important character sets for motorcycles like fuel guzzler-fuel efficient, sober-dynamic, etc. We will investigate this issue further.

Identities: Users have different identities-personal, institutional and social. In many cases, it depends on individual which identity he wants to bear and which he wants to be subsumed. The identity he wants to express through the artifact becomes the intended identity of the artifact. Today motorcycle styling is almost completely synonymous with identity of the rider. All three predominant classes of motorcycles have clear identities of the riders, particularly the identity of the self as he wants to describe himself as. Here, an interesting twist is that the motorcycle designer has to deal with two identities—one the rider actually possesses and the other, which he wants to express as. The following Table 8.1 illustrates the dichotomy:

Verbal metaphors: Verbal metaphors are important for human centered designers as they make as they make complex artifacts understandable and bring new perspectives in understanding them. Verbal metaphors are different from visual metaphors in that they have roots in literature, vocabulary, and cultural history. Visual imagery is evoked when we imagine motorcycles as horses, tigers, bulls, etc. But verbal metaphors use our imagination in more complex ways when we describe motorcycle as muscular, feminine, sexy, etc. Motorcycle does not have muscles, nor do they have anything to do with female creatures or sex. They words are rooted in our cultural memory and evoke certain images or characteristics which can be applied to motorcycles.

Table 8.1 Identities of motorcycle

Type of motorcycle	Actual identity	Expressed identity
Standard (means of transport)	Lower middle-income group	One who has arrived
	Conservative or liberal	Does not want to offend social sensibilities
Sports	Middle-high income group	Affluent, domination
	Conservative or liberal	Does not care about sensibilities
Cruiser (adventure/exploration)	High-income group	Classy, does not bother about class appearance
		Nature loving explorer in bare essentials

Narratives: Goldie [1] defines narrative as kind of representation of a sequence of events revealing certain coherence in what happened, configuring casual and other connections into a narratable episode or episodes. Humans live in stories and the objective of designing an artifact is that a coherent story with interfaces of humans with artifacts is feasible. Now there are innumerable ways how a motorcycle figures in human narratives. Best example is Che Guevara's "Motorcycle Diaries" where he describes his journey through Central American countryside and its poverty on a motorcycle. In this narrative, the motorcycle, the human being (Che and his friend Alberto), the geography of the countryside, and the social underpinnings get intertwined into a single narrative. Similarly, the movie "The Wild One" beautifully shows how the man, the machine, and the society interact with each other. As we all know, any story must have the following features:

1. Narrative smoothing: Creating a story by filling the missing gaps is called narrative smoothening. When we see a picture book or watch a movie, we should be able to imagine what happened between two subsequent frames. Then only the story becomes coherent.
2. Narrative sequencing: The sequence of events should be coherent. "He dialed a telephone number. The train blew up" and "the train blew up, he dialed a telephone number" have absolutely different connotations.
3. Narrative embedding: Keeping track of primary story while embedding other stories is called narrative embedding. In context of motorcycles, "changing the battery" is itself a complete story which can be embedded into another story of servicing the motorcycle.
4. Narrative schema: Artifacts should be designed in such a way that their interfaces should be narratable.

Consider, for example, the lines from "Motorcycle Diaries":

> Finally, at three in the afternoon, we left under the blazing sun that by the time we reached the sand dunes around the Medanos had become even hotter. The bike, with its badly distributed weight kept bounding out of control, wheels constantly spinning over. Alberto

fought a painful battle with the sand and insists he won. The only certainty is that we found ourselves resting comfortably in the sand six times before we finally made it out on to the flat.

Note the imagery between the lines and the interaction between the man and the machine, which make the narrative so complete. The moot point here for the designer is that, in the examples given above, the motorcycle has to complement the narrative and at no place the design should cause an interruption. The narrative of how a user experiences the artifact has many stages like when he feels the need, when he watches the artifact and when he undergoes the experience of using it. Different types of artifacts have different criticalities. Mostly, the narrative of the need is free from the design of the artifact like hunger, travel, etc. In case of abstract artifacts like poems this narrative has complex components like emotions. For even more complex and abstract artifacts like a social revolution, this narrative is more compelling, needs a detailed examination, and is subject to more intense questioning. For complex engineering objects, the second and third phase narratives are more important for the designer with the first phase narrative forming the essential backdrop.

Now that we have discussed the various meanings that a motorcycle can create for the user, what is important for a designer is to understand the engineering parameters of design which create these meanings. After observing many motorcycles, we found the following factors to be affecting the observer's emotion based on the context of the motorcycle, i.e., factors which affect the user due to the usage and its context:

8.1 Type: Cruiser, Sports, and Standard (Fig. 8.2)

The images that are conjured in our mind when we talk of cruiser or sports bikes are that of easy riders or bank robbers flying away on Hayabusas. The point is each of the bike categories has an image attached to it. And this image comes not only from movies and legends but essentially the engineering involved with it. Each of the categories has certain engineering characteristics attached to it and each characteristic results into a unique riding behavior and it is this riding behavior which forms the context. So here we will discuss the experiential narratives and the design attributes attached to the narratives.

Fig. 8.2 Cruiser, sports, and standard bikes

Experiential Narrative: Talking to different experienced riders, we derive the unique experiences from each category:

1. Standard bikes: Upright sitting posture with reasonably good stability and steering maneuverability, reasonably good speed with good load carrying capability.
2. Sports bikes: Aggressive posturing, good maneuverability with very high speeds, and thrill to match it.
3. Cruiser bikes: Relaxed posture with high stability at high speeds with a relaxed feeling at high speeds.

Visual Design Attributes: In fact, the main function of context is to set the context. This means preparing the rider for the future experience. So the visual attributes act as trailer for the real experience because they are easily relatable to the experiential narrative. We will discuss this issue specifically in the chapter on 'form'. Observing the various bikes, you get the visual design attributes for fuel tank surfaces as follows:

Standard bikes:

(a) Slow curvature on side
(b) Mild convexity
(c) No negative spaces
(d) Linear or curvilinear graphics.

Sports bike:

(a) Direction angular
(b) Distinct positive and negative spaces
(c) Top surface curvature has sharp break
(d) Sharp inter-space edges
(e) Minimal graphics with only the brand name.

Cruiser bikes:

(a) Axial direction horizontal
(b) No negative spaces
(c) Single side surface
(d) Minimal graphics
(e) Single surface on top.

Similarly, the visual design attributes for seat are:

Standard bikes:

(a) Direction horizontal
(b) Top surface flat with little step or very slow curvature
(c) Single rectangular or polygonal flat side surface
(d) Seat riding on to the fuel tank.

Fig. 8.3 Comparison of ergonomic triangles

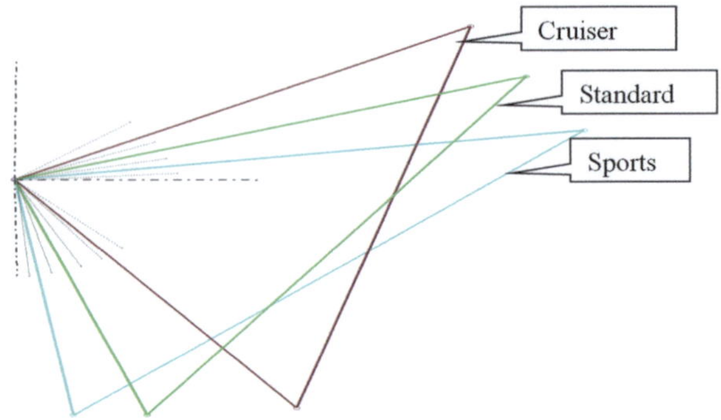

Sports bikes:

(a) Direction inclined
(b) Whole seat one single surface with fast curves
(c) Profile merging with fuel tanks.

Ergonomics:

Comparison of the three ergonomic triangles (Fig. 8.3) results into an interesting insight into the design attributes. The three different points of the triangles represent the rider seat (left), the handlebar grip (top right), and the rider foot rest (bottom).

Experiential Attributes: If we draw the max torque against the crankshaft RPM, we get a diagram (Fig. 8.4), which gives a very clear insight into the rider experience. (We have crankshaft rpm on *X* axis and peak torque (kgf-m) per 100 cc engine capacity on *Y* axis).

The sharp differentiation very clearly observed here is cruiser bikes like Harley Davidson FXDX, Indian Chief, and Honda VTX 1300 have their peak torques at very low rpms, whereas high powered sports bikes have peak torques at very high rpms. The first category is generally V twin engines with higher bore and stroke and relatively lower compression ratio and the basic objective is to provide high gradeability without lowering the gears to match with the experiential narrative of relaxed feeling at high speeds. The second-category bikes are generally four-cylinder engines with lower bore and stroke and higher compression ratio, to provide high torque at very high speed to have maximum power, to match with the experiential narrative of thrill and acceleration at high speeds. The bikes falling in the middle segment are single-cylinder commuter bikes. One interesting interloper is Harley Davidson VRod, which is a cruiser designed to give a sporty performance. Another interesting observation is that single-cylinder Indian cruiser bikes like Enticer and Eliminator are actually designed to perform like a standard bike but have the styling and ergonomics of a cruiser. Similarly, Indian sports bikes like

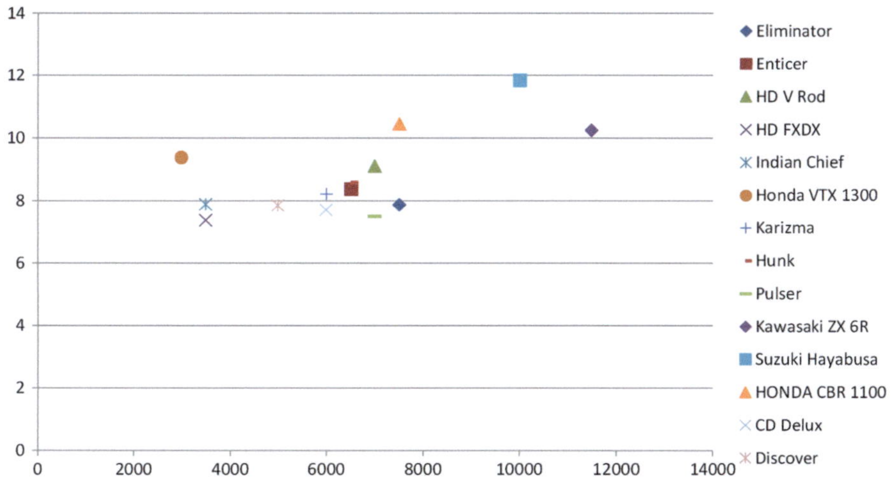

Fig. 8.4 Peak torques at crankshaft for motorcycle engines

Pulsar and Karizma also are having styling and ergonomics of sports bike but perform like a standard bike. The engine design parameters that are optimized to shift the peak torque are (Heywood):

1. Spark advance: The spark timing to give peak torques at different speeds is different. Timing is chosen for peak torque at desired speed.
2. Equivalence ratio: The carburetor or FI system is designed for fuel air ratio to give peak torque at desired speed.
3. Intake manifold length and diameter.
4. Bore stroke ratio: As the stroke increases, the lower in the rpm range, the peak torque will occur. In addition, a longer stroke engine will produce less peak HP but more peak torque.
5. Exhaust pipe length.

Rake angle and wheel base are the parameters which decide the stability and maneuverability of a bike. By looking at these values (refer following tables), it is clear that cruiser bikes need a higher rake angle because for these bikes stability is the most important requirement but maneuverability is low. Same goes with wheel base as higher wheel base provides more stability and lower turning radius (hence less maneuverability). Conversely, for sports bikes lower rake angles and lower wheel bases provide low stability but higher maneuverability. This can be explained by the fact that stability and maneuverability of a bike are broadly determined by two factors—yawing moment (stability decreases with increase) and steering ratio.

Yawing moment = twisting moment, self-aligning moment [2].

Self-aligning moment = a_t x cornering force where a_t = traila_t = a_{t0} (1 − λ/λ max) where a_{t0} = maximum trail, λ = side slip angle λmax = maximum side slip angle.

Since maximum trail is a function of rake angle, wheel size and front fork offset, largish rake provides maximum stability.

Now, steering ratio which indicates the steerability of the vehicle is given by expression

$$\xi = \frac{1}{1 - \left(\frac{1-k_{\psi r}}{k_{\lambda r}} - \frac{1-k_{\psi f}}{k_{\lambda f}}\right)\frac{V^2}{gp}},$$

where

ξ steering ratio
$k\psi r$ camber stiffness coefficient of front tire
$k\psi r$ camber stiffness coefficient of front tire
$k\lambda r$ cornering stiffness of front frame
$k\lambda r$ cornering stiffness of front frame
V vehicle velocity
P wheel base.

As per Cossalter, $\xi < 1$ means under-steering and $\xi > 1$ means over-steering. This means, smaller wheel base results in higher ξ hence better steerability.

Now, let us scan some actual data on rake angles, wheel sizes, and wheel bases and see, how these three parameters decide the fundamental character of a bike (Table 8.2).

You can very clearly see that lower rake angles (24–25°) have been selected for sports bikes for maneuverability, medium ones (26°) for standard and higher values (28–34°) for cruiser ones for stability. Now let us look at wheel base and rear wheel size of different types of motorcycles (Tables 8.3 and 8.4, respectively).

Here, longer wheel bases are favored for cruisers, and relatively lower ones for standard and sports bikes.

Table 8.2 Rake angles of motorcycle

Motorcycle	Rake
Suzuki Hayabusa	24°
Kawasaki ZX 6R	25°
HONDA CBR 1100	25°
Splendor	26°
CD Delux	26°
Discover	26°
HD FXDX	28°
Honda VTX 1300	32°
HD V Rod	34°
Indian Chief	34°

Table 8.3 Wheel base of motorcycles

Motorcycle	Wheel base (mm)
Discover	1305
Hunk	1325
Karizma	1355
Enticer	1380
Kawasaki ZX 6R	1400
Eliminator	1460
Suzuki Hayabusa	1485
HONDA CBR 1100	1490
HD FXDX	1622
Honda VTX 1300	1663
HD V Rod	1706.9
Indian Chief	1737

Table 8.4 Rear wheel size of motorcycles

Eliminator	130/90-15″
Enticer	120/90-16″
HD V Rod	240/40-18
HD FXDX	130/90-16
Indian Chief	150/90-16
Honda VTX 1300	170/80-15
Karizma	100/90-18
Hunk	100/90-18
Pulser	3 × 18
Kawasaki ZX 6R	180/55-17
Suzuki Hayabusa	190/50-17
HONDA CBR 1100	180/55-17

Here, we observe higher tire width with lower aspect ratio and relatively higher wheel diameters for sports bikes and lower width with higher aspect ratio and lower wheel diameters for cruiser bikes.

Now, how exactly these factor impact the emotion? The values of emotion scores observed after the survey can be referred in appendix. Here we can discuss the relationship between emotion and the aspects of context (construction, front facia, and wheels geometry).

8.2 Construction

The construction of motorcycles in general can be categorized on the basis of Bulk/lean and naked/ornamental (Fig. 8.5 shows three bikes showing these variations).

Fig. 8.5 Bikes varying in construction (*left*) standard, lean, and naked (*center*) cruiser, bulky, and naked (*right*) sports, bulky, and ornamental

Looking at the results in Fig. 8.6, the picture broadly confirms the general perceptions. Nothing to be surprised about as the picture has been created from perceptions only. It is the standard bikes which rule the sober emotion. And among the standard bikes, the bulky ornamental ones create perfect picture. Large speaking, glamor is an emotion which does not belong to any single-category. Sports bikes rule the dynamic emotion and intimidation can be generated by all three categories with different characteristics. Standard bikes create it with lean and naked construction, whereas cruisers can create it with bulky and naked construction and sport bikes can be equally effective with bulky and ornamental construction.

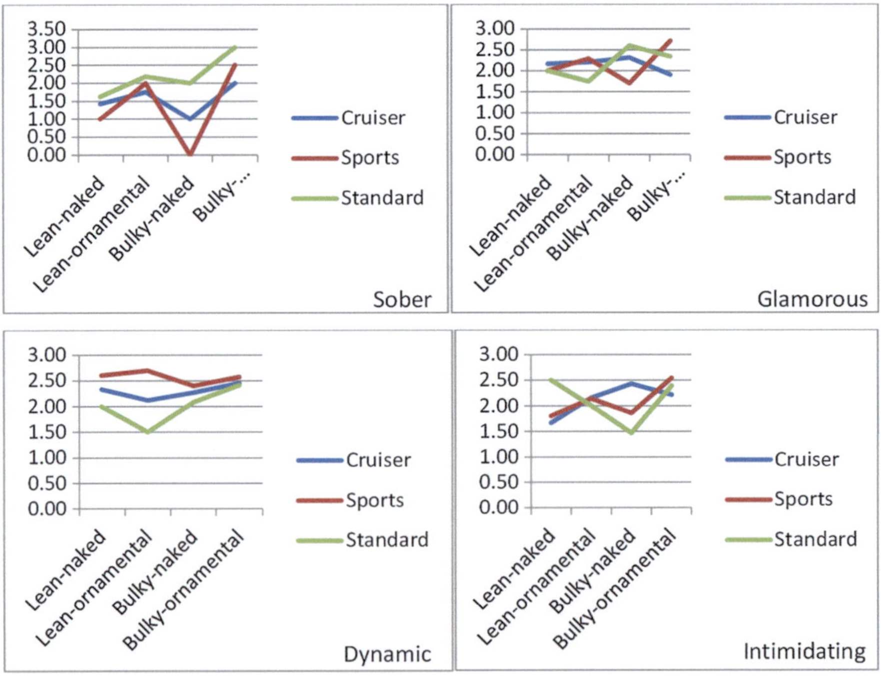

Fig. 8.6 Impact of construction on emotions

8.3 Front Facia

The front facia of a bike can be classified on the basis of presence of visors (Fig. 8.7), shape of headlights (Fig. 8.8) and whether visors are fitted with windscreens (Fig. 8.9). Let us have a look.

Once we plot the emotion scores based on the classifications (Fig. 8.10), we get the curves which explain the relationship.

In cannot get clearer than this. Round headlights are best for sober styling with rectangular ones a close second and trapezoidal ones are the ones to go for glamor, dynamism or intimidation, and so far as dynamism is concerned trapezoidal is by far the only choice. Again, for sobriety, without visor is the best option as validated by Bullet from Royal Enfield and the stylists can go with visors also if combined with round and rectangular headlights. Glamor appeal is best achieved by combining trapezoidal or rectangular head light with visor without windshield whereas windshields should be preferred for dynamism and intimidation.

8.4 Wheels

We have previously observed that wheels can vary between big and small (Fig. 8.13), cast and spoke (Fig. 8.12) and wide and narrow (Fig. 8.11).

Looking at the results plotted (Fig. 8.14); by and large cast wheels are the choice, whether you are looking for sobriety, glamor, or dynamism. But when it comes to intimidation, spoke wheels come out of hibernation it seems. Though, intimidation is at its best for big narrow wheels, where cast wheels are equally

Fig. 8.7 With and without visors

Fig. 8.8 Varying shapes of headlights (*left*) circular (*center*) rectangular (*right*) trapezoidal

Fig. 8.9 Only visor and visor with windscreen

effective in creating the overpowering feel. If you talk of the difference between narrow and wide wheels as a trend, there is no such trend and both have almost equal capability to create a desired emotion with one or the other choices for big or small and cast or spoke. So far as the trend between big and small wheels is concerned, the trend is ambivalent for sober and glamorous, smaller wheels are preferred for dynamism due to their perceived capability to create maneuverability and big wheels are preferred for their perceived stability, their long association with cruisers and of course simply because of their bigness.

Making sense of a meaning of design of bikes is the context which converges the social, emotional, and engineering in one single whole. What we design not only fulfills its engineering need but it also generates a vocabulary in the mind of the user out of his experiential and social narratives and it is this vocabulary which ultimately results in the emotion which the designer desires. Ultimately, we will see

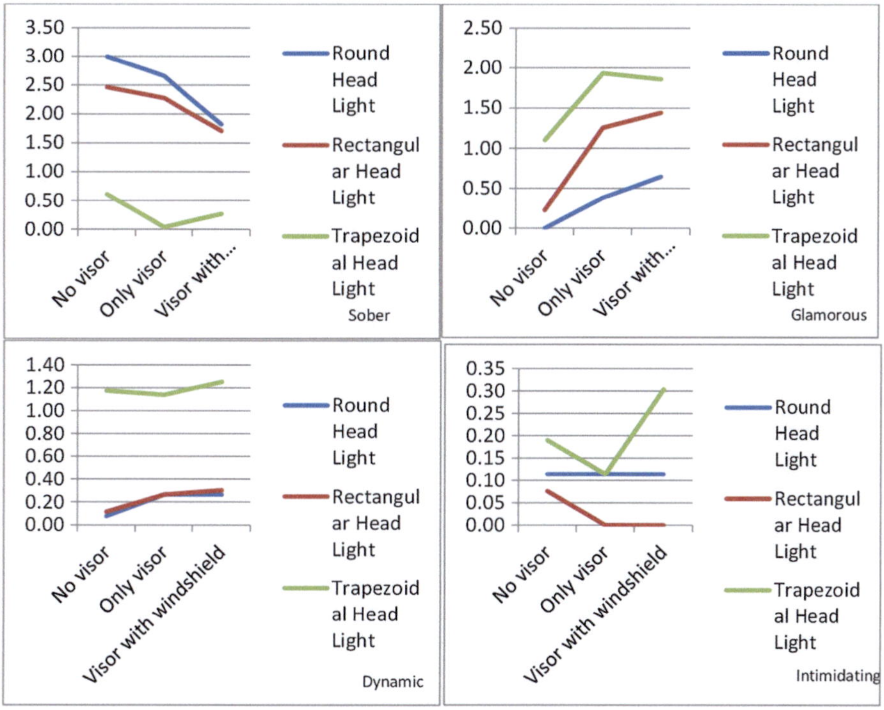

Fig. 8.10 Impact of visors and windscreens on emotions

Fig. 8.11 Wide and narrow tires

Fig. 8.12 Spoke and cast wheel

Fig. 8.13 Big and small tires

that it is this perception germinating from the context which dominates the overall process of deciding the emotion. In fact, the whole contribution of engineering in this process is through this context whether it is the power of the engine, the comfort provided by the seat, or the drivability provided by the overall geometry and the suspensions. This is the final takeaway for the hardcore engineers—that, engineering is nothing but sociology through science.

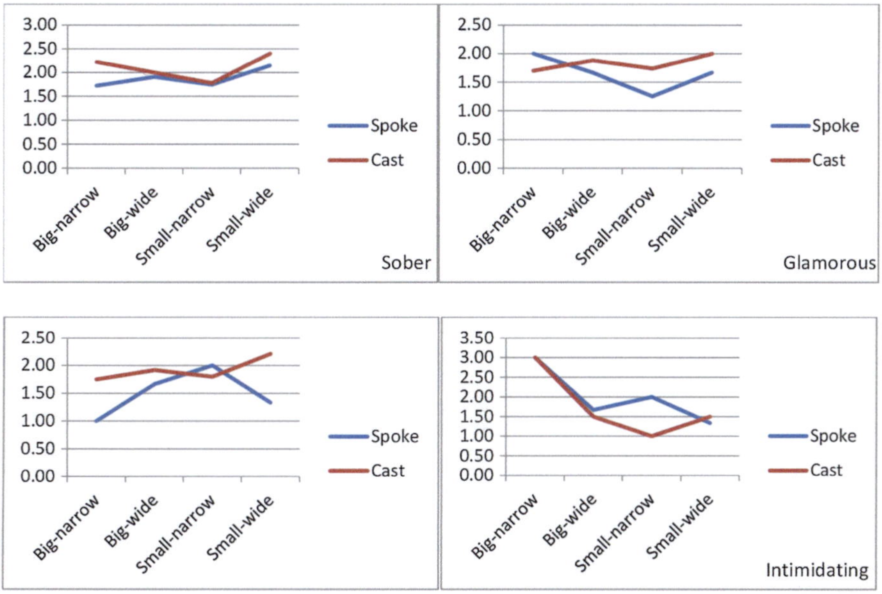

Fig. 8.14 Impact of wheels on emotions

Appendix 1: Emotion Scores for Motorcycle Types

		Cruiser				Sports				Standard			
Lean	Naked	S	G	D	1	S	G	D	I	S	G	D	I
		0.46	0.60	1.94	1.16	0.09	0.74	2.41	1.25	3.00	0.09	0.14	0.23

	Ornamental	S	G	D	I	S	G	D	I	S	G	D	I
		0.32	0.51	1.67	1.90	0.14	0.74	2.87	1.39	1.11	2.50	0.14	0.09

Bulky	Naked	S	G	D	I	S	G	D	I	S	G	D	I
		0.19	0.88	2.27	1.43	0.00	0.79	2.22	1.20	0.19	0.60	2.31	1.02

	Ornamental	S	G	D	I	S	G	D	I	S	G	D	I
		0.46	0.60	1.94	1.16	0.23	0.88	1.67	2.59	0.28	0.32	1.90	1.54

Appendix 2: Emotion Scores for Wheels

<table>
<tr><td></td><td></td><td>SPOKE</td><td>CAST</td></tr>
<tr><td rowspan="2">BIG WHEEL</td><td>NARROW TYRE</td><td>

S	G	D	I
3.00	0.08	0.04	0.12

</td><td>

S	G	D	I
0.81	1.58	0.85	0.12

</td></tr>
<tr><td>WIDE TYRE</td><td>

S	G	D	I
2.55	0.41	0.20	0.20

</td><td>

S	G	D	I
0.32	1.90	0.93	0.24

</td></tr>
</table>

<table>
<tr><td></td><td></td><td></td><td>SPOKE</td><td>CAST</td></tr>
<tr><td>3</td><td rowspan="2">SMALL WHEEL</td><td>NARROW TYRE</td><td>

S	G	D	I
2.75	0.20	0.08	0.08

</td><td>

S	G	D	I
0.65	1.62	0.73	0.12

</td></tr>
<tr><td>4</td><td>WIDE TYRE</td><td>

S	G	D	I
2.35	0.81	0.16	0.16

</td><td>

S	G	D	I
0.49	1.13	1.70	0.36

</td></tr>
</table>

Appendix 3: Emotion Scores for Front Facia

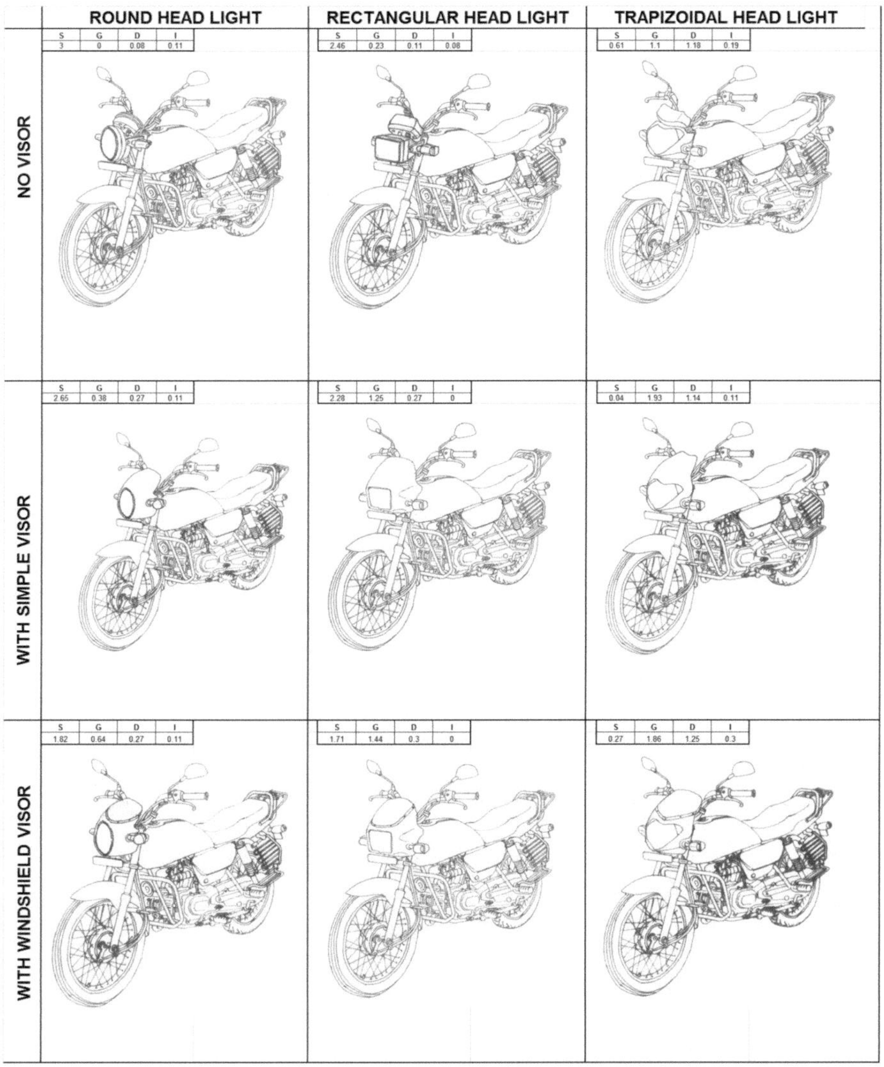

References

1. Goldie, Peter. 2009. Narrative thinking, emotion and planning. *The Journal of Aesthetics and Art Criticism* 66(2): 97–106 (Winter).
2. Cossalter, Vittore. 2002. Motorcycle Dynamics. Race Dynamics, Greendale, WI, USA, 2: 57–59, 3: 108.

Chapter 9
The Shape of a Motorcycle

The hands of a man were not made 700,000 years ago to design cars or bottles or who knows what. They were made to touch the beautiful shapes of a woman's body. These hands cannot forget the breasts of the woman they touched the night before. If I take a material, it's for me the body. I still have the woman's warmth in my hands.

Thus spoke Luigi Colani, the founder of London Design Museum, a vehemently sensuous supporter of "form follows emotion" school of design. "Form follows function" versus "form follows emotion" has been the subject of an eternal debate in the world of design. Though, today is the age of functional design thought from Bouhaus to the Ulm school of design, it is not only the Michelangelo's David but the body of Beetle (the Car) as well, which has to appeal to the emotional part of the brain.

The body of a motorcycle is a more complicated story though. To the eye of an observer, the body of Beetle is one single entity, but motorcycle is a bunch of parts. Though all parts constitute the whole, not all of them appeal to the emotions in the same way. These parts can be largely divided into two categories—those having geometric form and those having organic forms. The parts with geometric form are those which form the engineering architecture of the bike like engines, wheels, brakes, etc. It is not that these parts do not appeal to emotions. They certainly do but they appeal in a contextual way and this has been discussed in the chapter on context. The physical sensuality of a bike (the way Luigi Colani imagines) is dictated by parts having organic form, which are styled by studio designers. Even here, we are not discussing some of these parts like lights, wheel rims, and grip as they are more of functional parts. More importantly they are not part of the whole called body of motorcycle. Moreover, there is still another category of parts, which actually form the part of motorcycle body like the front and rear fenders, but they are more functional and their visibility is too little to affect the emotional appeal of a bike. So, for our purpose we will be discussing only those parts which constitute the body and are sufficiently visible. The left out organic parts definitely affect but not to the extent that the overall emotional appeal is altered. These parts are the following:

S. Chandra, *Aesthetics: Quantification and Deconstruction*,
Design Science and Innovation, https://doi.org/10.1007/978-981-10-6235-3_9

a. Fuel tank
b. Side covers
c. Visor
d. Seat

The illustration in Fig. 9.1 shows these parts and you can realize, how they dictate the look of the bike. In this chapter, we will be discussing the various possibilities for the form of these parts and their impact on the emotion created by motorcycles.

9.1 Fuel Tank

Let me make an honest confession. After spending almost whole of my life with design of bikes, I am still mesmerized by the magic of fuel tanks. Like the theme song of a movie, it declares the character of the bike in one single stroke. More precisely, it can be compared to the breasts of a woman. Though, their main function is to store and provide milk to the newborn, they actually act as the centerpiece of attraction for the female physicality. Similarly, though the fuel tank's main function is to store and provide fuel for the bike, they actually act as the centerpiece of bike's physical attraction. Take any cult bike—the Harley Iron series (see Fig. 9.2, left), the Ducati Monster series, Royal Enfield Bullet series (see Fig. 9.2, right), or Bajaj Avenger or Pulsar, it is the fuel tank which acts as your gateway to know the bike. Of course, we are talking about step-over bikes only as step through bike have their fuel tanks hidden below the seat and precise for this reason are mostly considered poor cousins of better known step-overs.

Fig. 9.1 Parts of a motorcycle dictating its form

Fig. 9.2 Harley iron (*left*) and Enfield bullet (*right*)

Though there is nothing in the Holy Bible or law of any country which restricts the shape of a step-over fuel tank, we can safely make some generalizations based on observation, and this is an observation which holds true since the beginning of motorcycle history till now. Almost all fuel tanks have three sides, which are visually available—left, right and the top side, where left and right sides are mirror images of each other. Hence practically we are left with two faces. What we have done here for our study is that we created the various possibilities though digital modeling and collated their impact on human emotions statistically. Based on observation of various models worldwide, following possible considerations for categorization emerge:

1. Fillets: Fillets are the most visible differentiators which build and identify the character of not only the fuel tank but the bike as well. These fillets can be characterized as sharp, medium, and large as shown in Figs. 9.3 and 9.4.
2. Surfaces: The sides and the top of a fuel tank are generated by combinations of surfaces which are either flat, convex, or concave as illustrated in Figs. 9.5 and 9.6.
3. Space combinations: Like a human body, the chief determinant of the character of the body is the combination of spaces which is either positive/positive or positive/negative. A negative/negative combination is not considered as it is highly unlikely (see Figs. 9.7 and 9.8).

Fig. 9.3 Fuel tank fillets shapes. Small (*left*), medium (*center*), and large (*right*)

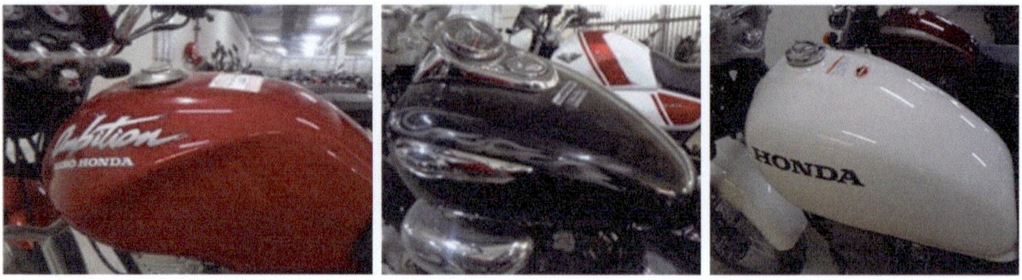

Fig. 9.4 Fuel tank with varying fillets. Small (*left*), medium (*center*), and large (*right*)

Fig. 9.5 Fuel tank surface variations (side surface) Flat (*left*), convex (*center*), and concave (*right*)

Fig. 9.6 Fuel tank with varying side surfaces. Flat (*left*), convex (*center*), and concave (*right*)

4. Curvature: Curvatures with a higher rate of change are known as fast curves and those with low rate of change are slow ones (see Figs. 9.9 and 9.10).

Based on the generalizations we have just discussed, the form of a fuel tank, fundamentally has three elements: the top geometry, the side geometry, and the fillets. To further restrict the scope of discussion, I propose to ignore the top geometry because the variation in the top geometry is very little. So, now we are left with the task to study the impact of side geometry and fillet. The juice of the story lies here, where we collected the emotion score from respondents and collated them graphically. The trends for emotions tell their own stories.

Fig. 9.7 Fuel tank space combinations. Positive–positive (*left*) and positive–negative (*right*)

Fig. 9.8 Fuel tank with varying space combinations. Positive–positive (*Left*) and positive–negative (*right*)

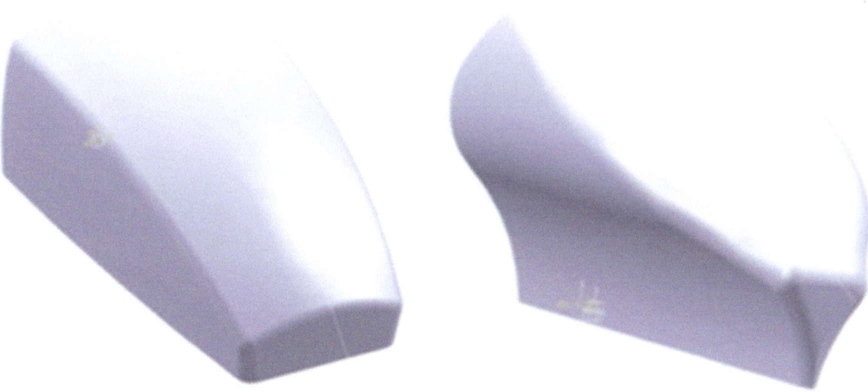

Fig. 9.9 Fuel tank space curvatures. Slow (*left*) and fast (*right*)

Fig. 9.10 Fuel tank with varying curvature. Fast (*left*) and slow (*right*)

Sober: One thing that comes out loud and clear (Fig. 9.11) is that flat and mild convex surfaces are best for generating sober emotions. This works best with medium fillets. Medium fillet is also very effective when convex surfaces combine with flat surface. But when the fillets have to deal with either combination of convex surfaces or convex and concave surfaces, it is the large fillets which fit the bill. Another general trend is the preference for slow curves, so far as sober emotions are concerned.

Looking at the individual scores, the fuel tank which evokes the best, rather perfect sober emotion is one with mild fillets and mild convex sides. The perfect example for this is the fuel tank for Hero Splendor (see Fig. 9.12).

Glamor: The plot (Fig. 9.13) looks complicated but a slightly deeper glance makes it clear that a combination of convex and concave surfaces is the way to go for creating glamor in fuel tank. Whether the fillets are large or medium or sharp, the combination of convex and concave surface can work the magic and this magic works both with fast and slow curvatures. If you are stuck with combination of

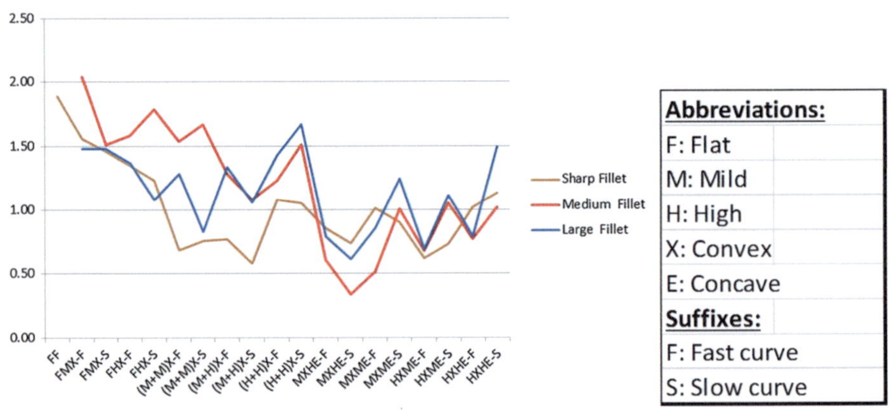

Fig. 9.11 Impact of fuel tank form on sober emotion

Fig. 9.12 Fuel tank with best sober emotion (Hero Splendor)

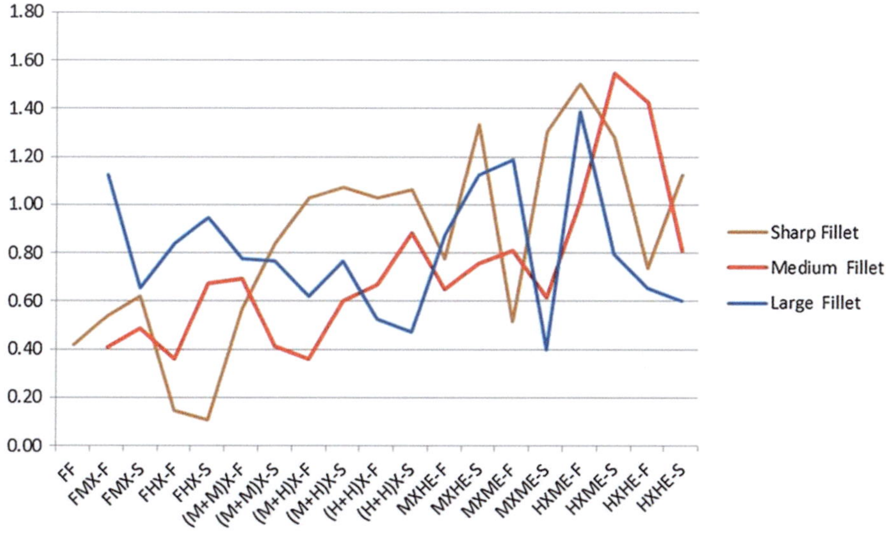

Fig. 9.13 Impact of fuel tank form on glamor emotion

convex surfaces, you must work with sharp fillets but if you are left with no option but flat and mild convex surfaces, large fillets should be preferred.

The best example of convex and concave combination with a large or medium fillet creating a glamorous output is the fuel tank of Hero Glamour (Fig. 9.14). Though the name of the bike itself gives away the game, it is actually the fuel tank shape which justifies the name.

Dynamic: The message is clear (Fig. 9.15). Combining convex and concave surfaces with medium fillets are the best way to generate dynamism in fuel tank design. One can work with combination of convex surfaces as well for generating a lesser degree of dynamism but with sharp fillets. The best examples of this are Pulsar fuel tanks (Fig. 9.16).

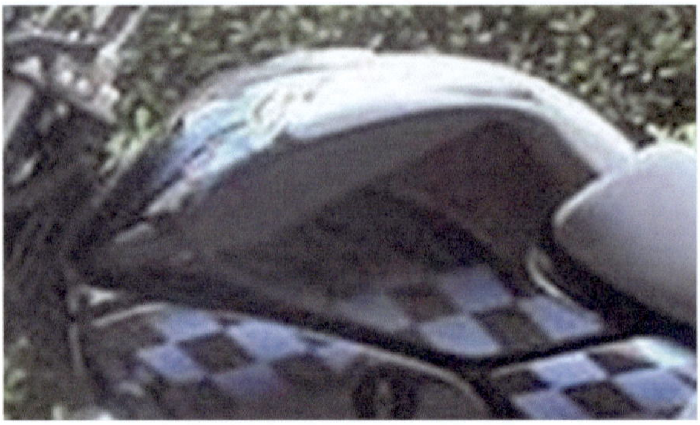

Fig. 9.14 Fuel tank with best glamor emotion

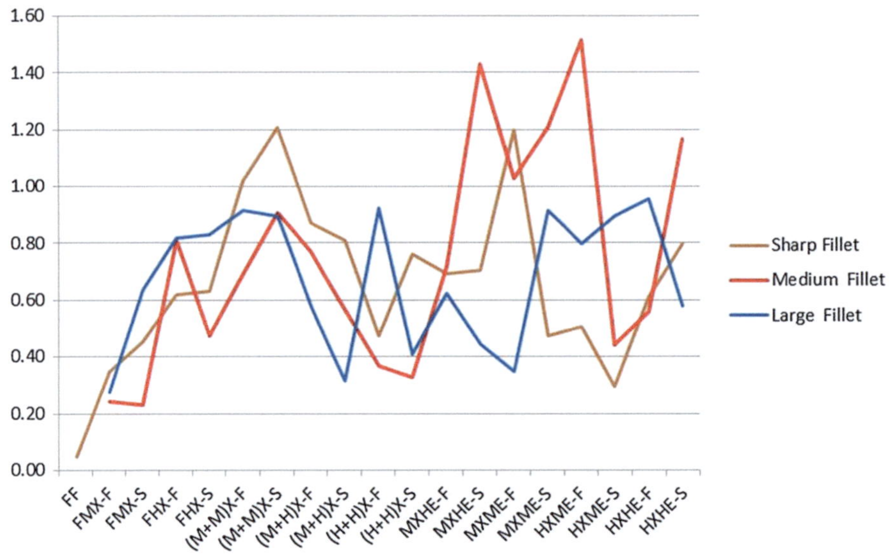

Fig. 9.15 Impact of fuel tank form on dynamic emotion

Fig. 9.16 Fuel tank with best dynamic emotion

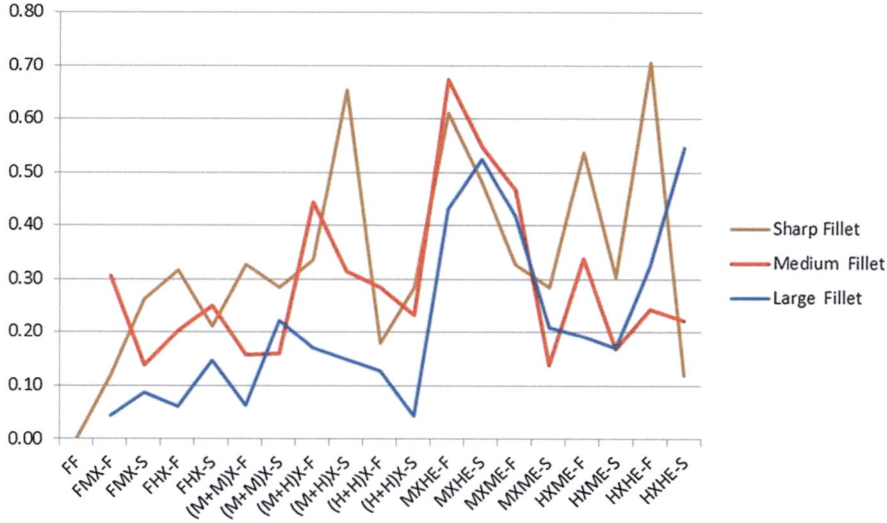

Fig. 9.17 Impact of fuel tank form on intimidating emotion

Fig. 9.18 Intimidating emotion in KTM Duke

Intimidating: The plot (Fig. 9.17) makes it clear that sharp fillets and fast curves are the undisputed choice which ever combination you are working with. If generating intimidation is even remotely desired, one should not touch flat surfaces with a barge-pole and combination of convex and concave surfaces are most preferable shapes. One look at the fuel tanks of KTM Duke (Fig. 9.18) will clear any doubt on this.

9.2 Side Cover

Side covers are parts which are used to cover various functional parts below the frame. It has seen many a metamorphosis through the history. In the initial years, it was almost nonexistent and made its appearance in mid of the last century. Slowly as the styling of bikes evolved into various visual architectures, the presence of the side cover became more and more prominent.

Generally, only one face of the cover is visible with side edges all over the periphery. By observing various motorcycles, following classifications emerge:

1. Shape: Triangular, polygonal (Fig. 9.19)
2. Spaces: Positive, Positive/negative (Fig. 9.20)
3. Fillets: Sharp, large (Fig. 9.21)
4. Surfaces: Flat, convex, flat step, flat/flat angle, flat concave, flat convex, convex-concave (see Fig. 9.22).

The trends based on survey comprising all possible combination of all the above classification have been explained in the following discussions.

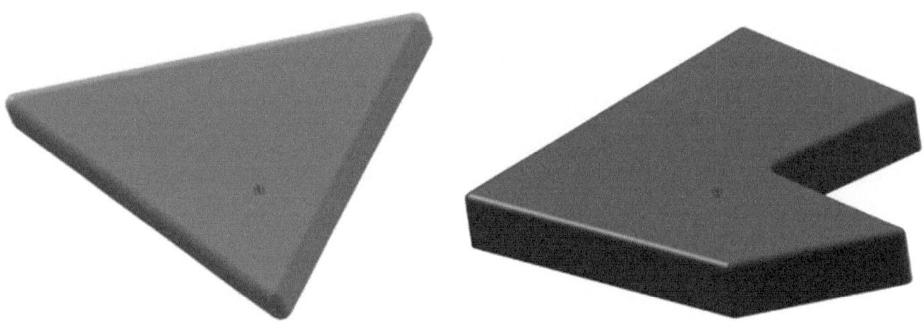

Fig. 9.19 Side cover shapes. Triangular (*left*) and polygonal (*right*)

Fig. 9.20 Side cover spaces. Positive (*left*) and positive–negative (*right*)

Fig. 9.21 Side cover fillets. Sharp (*left*) and large (*right*)

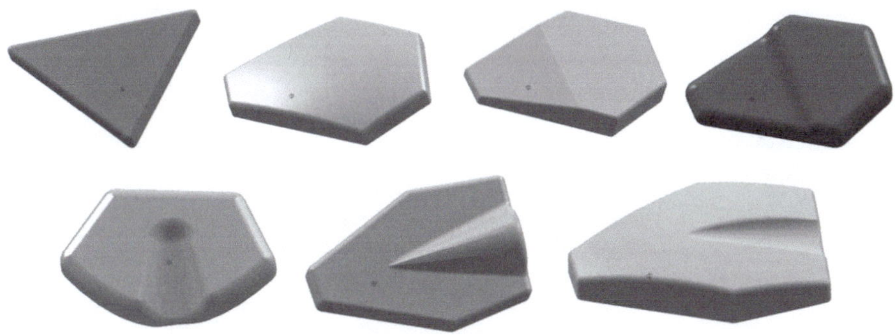

Fig. 9.22 Side cover surfaces. Flat (*top row first*), convex (*top row second*), flat-step (*top row third*), flat/flat-angle (*top row last*), flat-concave (*bottom row first*), flat-convex (*bottom row middle*), and convex-concave (*bottom row last*)

Sober: One thing is very obvious from the plot in Fig. 9.23. To get the best score for sober emotion, triangular shapes with large fillets are the best option. This is exemplified with Bullet (Fig. 9.24, left) and Splendor side covers (Fig. 9.24, right). But triangular shapes are good option only for singular surfaces like flat or convex surfaces. On the other hand, sharp fillets need not be avoided altogether and they can be a good option with recto-curvilinear shapes with negative intrusions. To be fair, it will be wrong to generalize either sharp or large fillets to be suitable for sober effect.

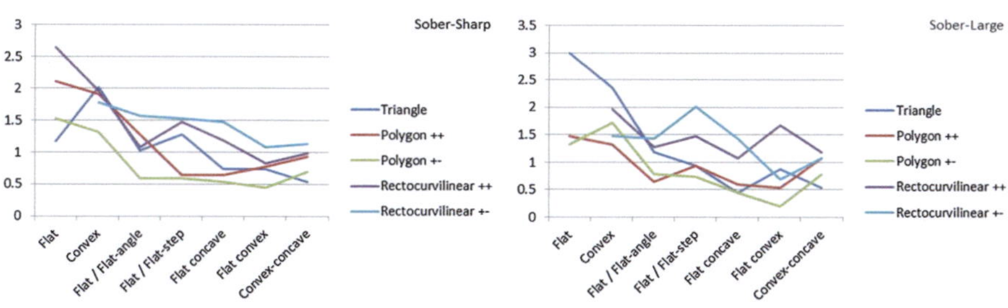

Fig. 9.23 Impact of side cover shape on sober emotion

Fig. 9.24 Side covers with best sober emotion

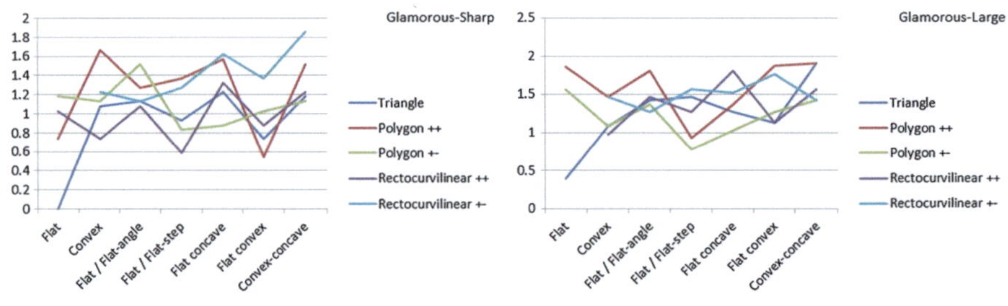

Fig. 9.25 Impact of side cover shape on glamor emotion

Glamorous: The plots for glamor emotion in Fig. 9.25 show that neither sharp nor large fillets can be generally recommended for glamor effect. With sharp fillets, you can get good glamor quotient with convex or convex-concave space combinations by using polygonal or recto-curvilinear shapes, respectively. On the other hand, large fillets can also be used with flat, flat and inclined combination, or convex–concave combination with polygonal and recto-curvilinear shapes, respectively. On the whole, this emotion offers a lot of choices to the designers.

The individual scores for glamor emotion indicate the best score for large fillets with polygonal shapes and this is vindicated by the side covers of Passion Plus by Hero (Fig. 9.26).

Dynamic: This emotion offers clear choices (see Fig. 9.27). Singular surfaces need to be avoided at all cost and space combinations have to be used to generate dynamism. In general, sharp fillets provide an edge and create a magic with polygons and this is clear if you look at the side covers of Pulsar series of bikes by Bajaj (Fig. 9.28). On the other hand, large fillets can also be deployed carefully by using recto-curvilinear shapes and flat concave space combinations Overall, this emotion provides a direction for space combinations but offers wide options for shapes and narrow ones for fillets.

Fig. 9.26 Side cover with best glamor emotion

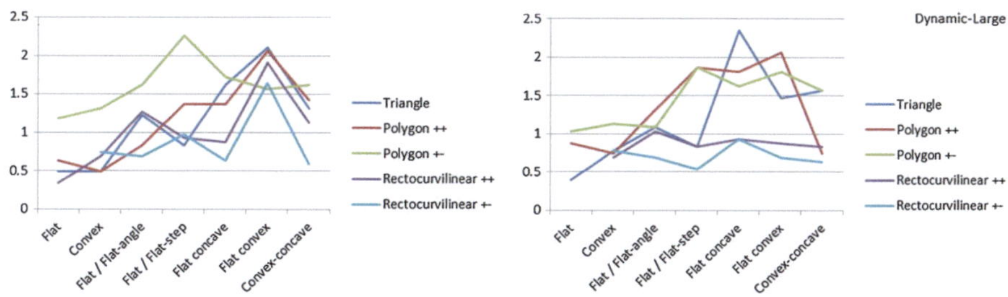

Fig. 9.27 Impact of side cover shape on dynamism emotion

Fig. 9.28 Side cover with best dynamism emotion

Intimidating: Again, the graphs in Fig. 9.29 clearly show that singular surfaces are to be strictly avoided and polygons are the clear choice both for large as well as sharp fillets, though sharp fillet has an edge. In the case of sharp fillet, flat convex space combination is to be preferred as exemplified by KTM Duke covers (refer Fig. 9.18), whereas flat and flat-step space combination is the way to go for large fillets.

Till now, we have discussed the trends separately. Some bullet points which come out of these discussions are here for you to take away:

a. In general, flat surfaces either with triangular or recto-curvilinear profile generates a sober character.
b. Large filleted polygons either with flat surfaces or flat/flat angles generate glamor.

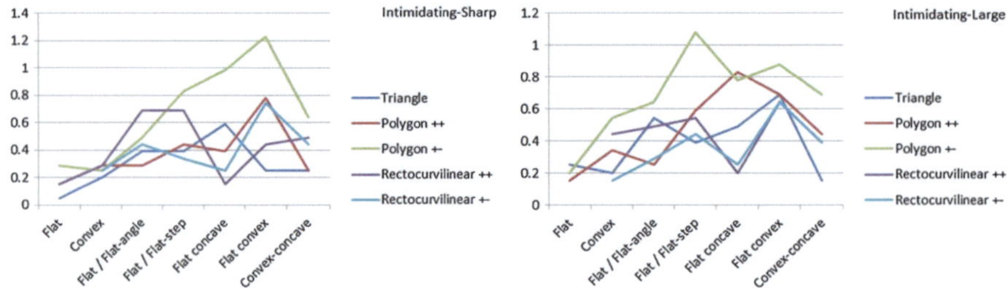

Fig. 9.29 Impact of side cover shape on intimidating emotion

c. Sharp fillet polygons or triangles either with flat/flat angle surfaces or flat convex surfaces create dynamism.
d. Element of intimidation generated by any form of side cover is very low compared to other characters. In some cases, it can be a supplementary character as discussed in point "g".
e. Convex surfaces either with polygon or recto-curvilinear profile generate a combination of sober and glamorous characters.
f. Flat/flat angle surfaces or convex-concave surfaces with triangular or polygonal profile generate a combination of glamorous and dynamic characters.
g. Sharp fillet polygons with flat convex surfaces generate a dynamic character supplemented by an element of intimidation.

9.3 Seat

One point which you will definitely agree is that I do not have to illustrate the seat of a motorcycle by a photograph. Whether it is Brando and Murphy Mary in "The Wild One" or Rishi and Dimple in "Bobby," cinematic legends have been created on this part of the motorcycle. Needless to say, it is an important part which defines the visual character of a motorcycle. Coming to the geometry of the seat, it has two important components—top face and the side face.

Top face:

Single edge: Flat, inclined, and curved with straight edge and curved edge as shown in Fig. 9.30.

Fig. 9.30 Seats with single edge top face. Flat with straight edge (*first*), flat with curved edge (*second*), inclined with straight edge (*third*) and inclined with curved edge (*last*)

Polygonal top edge: Stepped, inclined-flat, flat curve, and inclined-curved (see Fig. 9.31).

Side face:

Flat bottom, inclined bottom, stepped bottom (Fig. 9.32).

Now let us see how these variables impact the emotions generated by the seat.

Sober: The best case option for a sober design is inclined seat with a flat bottom as clearly shown in Fig. 9.33. But seat is a part which has a highly functional profile and since designers are constrained by functional requirements, they cannot always opt for the best case option. The other options available for a sober emotion are inclined-flat top with a curved bottom. The general trend coming out from the trends is that curved profiles are avoided for sober emotion.

Glamorous: On the other hand, top profiles with curves combining with flat or inclined and steps are preferable for a good glamor appeal (Fig. 9.34). When it comes to choosing between shallow and deep curves, the choice is ambivalent and

Fig. 9.31 Seats with polygonal edge top face. Stepped (*first*), inclined-flat (*second*), flat-curved (*third*), and inclined-curved (*last*)

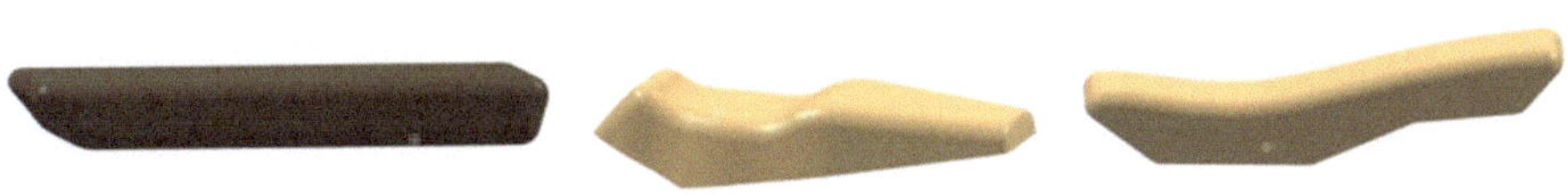

Fig. 9.32 Seats with varying side face. Flat bottom (*first*), inclined bottom (*second*), and stepped bottom (*last*)

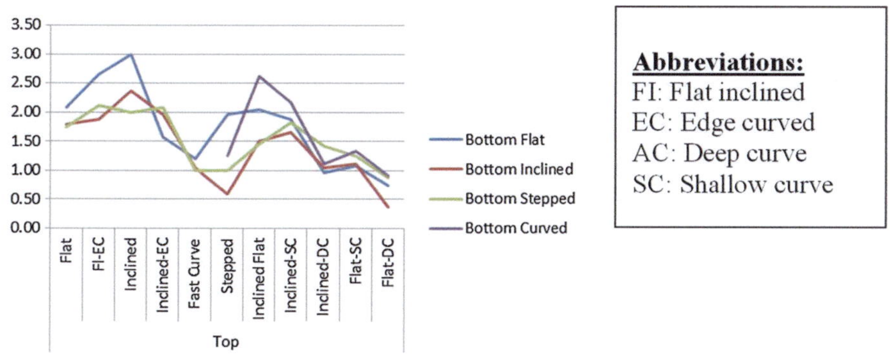

Fig. 9.33 Impact of seat shape on sober emotion

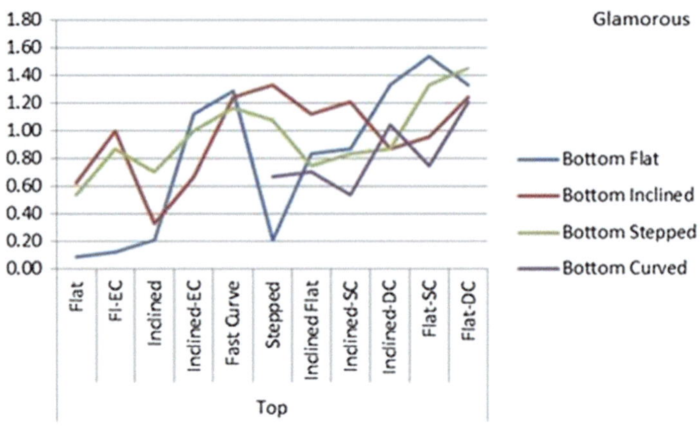

Fig. 9.34 Impact of seat shape on glamor emotion

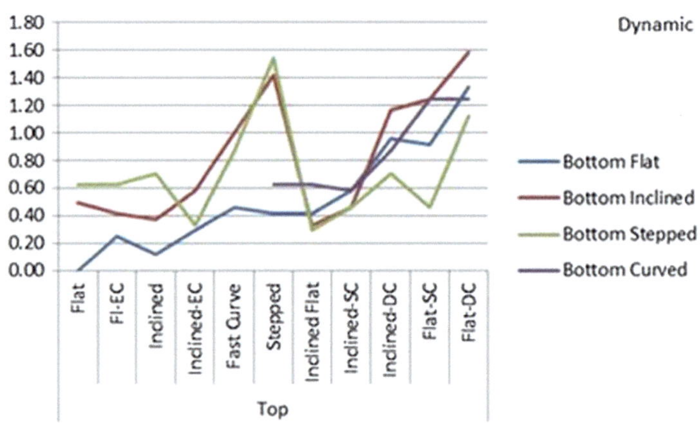

Fig. 9.35 Impact of seat shape on dynamism emotion

depends on the bottom profile you choose. And when it comes to choosing the bottom profile, again, flat bottoms work fine with curved top profiles and inclined bottom with fast curves, steps, or inclines with flat or shallow curves.

Dynamic: Again, the general trend is in favor of curves (Fig. 9.35). Top profiles with steps and flat with deep curves are best choices for generating dynamism. One clear recommendation is to avoid singular top profiles. Another clear choice is to go for deep curves instead of shallow ones. So far as the choice of bottom profiles are concerned, it depends on the top profile you choose.

Intimidating: One sweeping generalization can be made without much ado. If you want to generate an intimidating emotion (Fig. 9.36), seat cannot contribute to this and the whole analysis amounts to differentiating between miniscule amounts of intimidation. Still, relatively speaking, stepped bottom and inclined with deep curve is the preferred option.

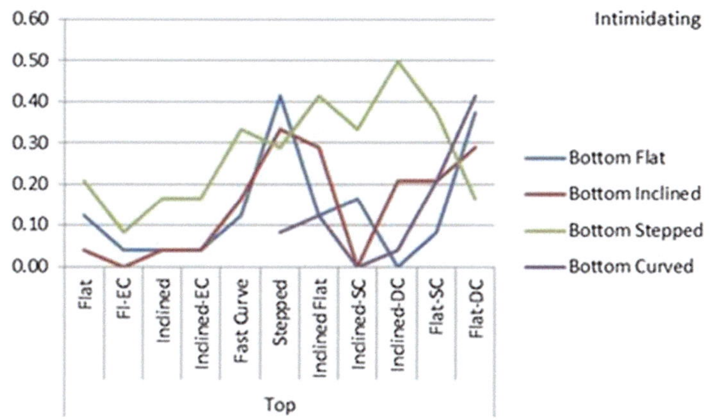

Fig. 9.36 Impact of seat shape on intimidating emotion

General trends observed in this survey are as follows:

a. Flat or inclined top with flat bottom generates a sober character.
b. Seats with dual surface top (either flat and curved top or stepped top) and stepped and bottom generate glamorous or dynamic characters or both.

9.4 Visor

Visor made a very late appearance on the stage. It made its first appearance in shape of windscreens, which were generally fitted as accessories by users. But windscreen had their own aerodynamic issues and a solution was finally found in shape of visor in the dying stages of twentieth century. So it happens to be the youngest member of the family.

Visor form has three basic elements—top contour, headlight shape, and surface.

Top contour: Flat, convex, and concave (Fig. 9.37).

Headlight shape: Rectangular and trapezoidal (Fig. 9.38).

Surface: Flat, convex, flat-concave, flat-convex, convex-concave, and convex-convex (Fig. 9.39).

Fillets: Sharp fillet and large fillet (Fig. 9.40).

Based on these classifications, the trends for impact of these factors on the emotions have been discussed as follows.

Sober: Surface combination with flat and concave surfaces with sharp fillets gives the best results for sober emotions (refer Fig. 9.41). Flat surfaces with sharp fillets also create the same effect, albeit to a lesser extent. The headlight profile and contours do not make a drastic impact as far as these combinations are concerned. Large fillets are effective only in case of flat contours.

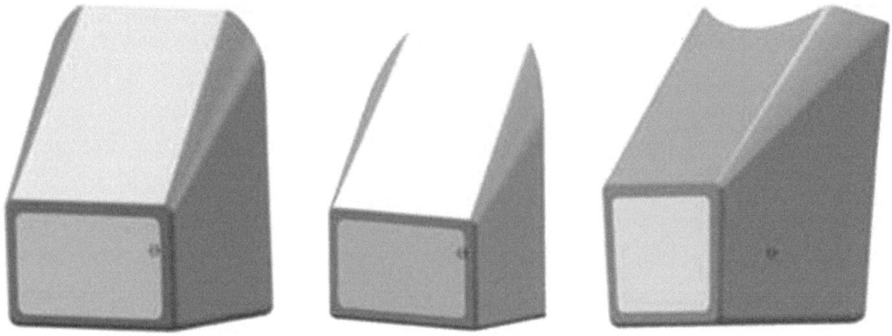

Fig. 9.37 Visors with varying top contours. Flat (*left*), convex (*middle*), and concave (*right*)

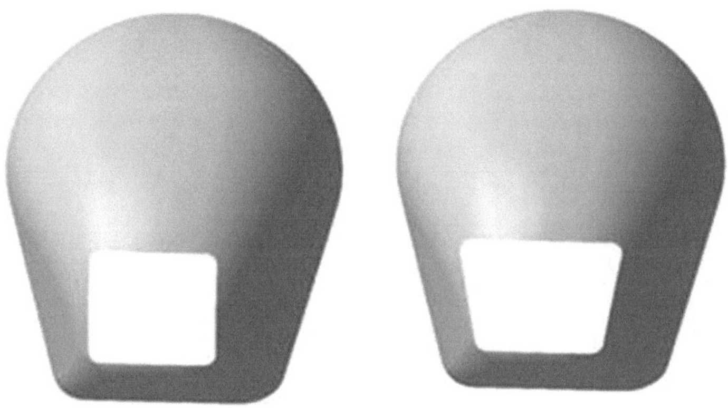

Fig. 9.38 Visors with varying headlight shape. Rectangular (*left*) and trapezoidal (*right*)

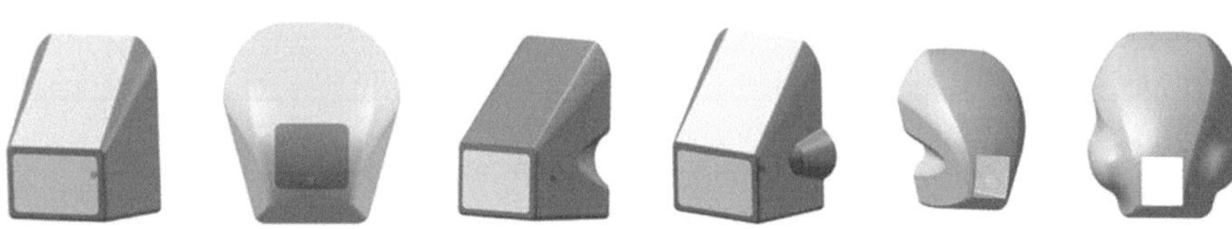

Fig. 9.39 Visors with varying surfaces. *In order from left to right* Flat, convex, flat concave, flat convex, convex-concave, and convex-convex

Glamorous: Convex-concave surface combination creates relatively better glamor appeal with sharp fillets and concave contours and rectangular head lights (Fig. 9.42). Convex surfaces with large fillets, convex contour and rectangular headlights generate almost the same impact. But overall, you cannot get the same impact as in the case of sober designs where the score goes up to 3.

Dynamic: Sharp fillets are definitely better if you want to create dynamic visor design and convex-concave surface combination is the preferred option (Fig. 9.43).

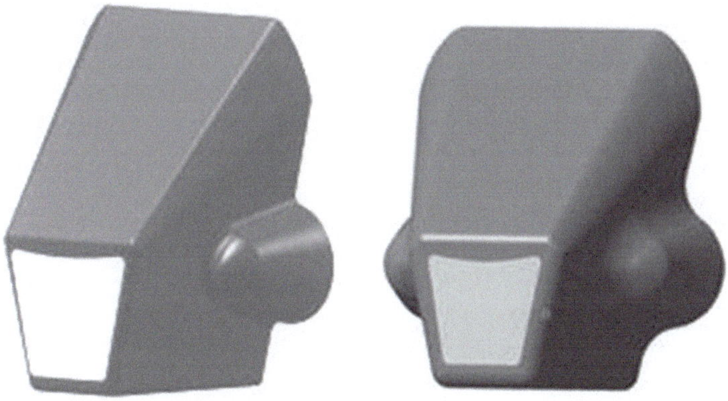

Fig. 9.40 Visors with varying fillets Harp fillet (*left*) and large fillet (*right*)

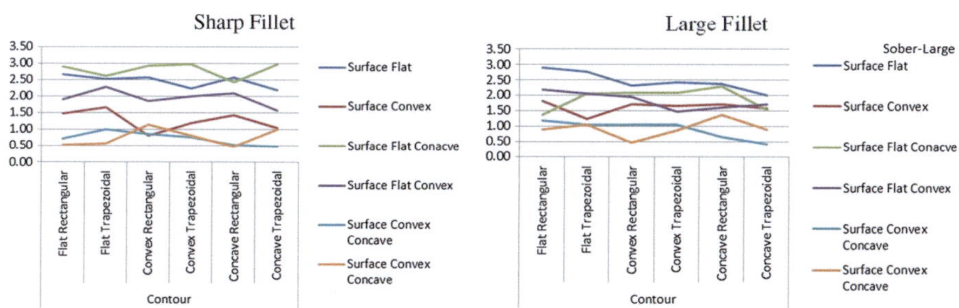

Fig. 9.41 Impact of visor shape on sober emotion

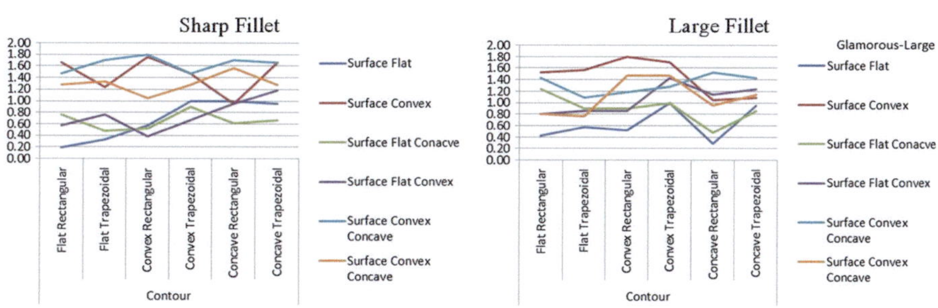

Fig. 9.42 Impact of visor shape on glamor emotion

With convex-concave combination, even large fillets can work with concave contours and trapezoidal headlights. But in overall context, the scores generated are not comparable with sober designs.

Intimidating: In general, as Fig. 9.44 shows, sharp fillets are more capable of generating intimidation except for a one-off exception. Again, this is in limited

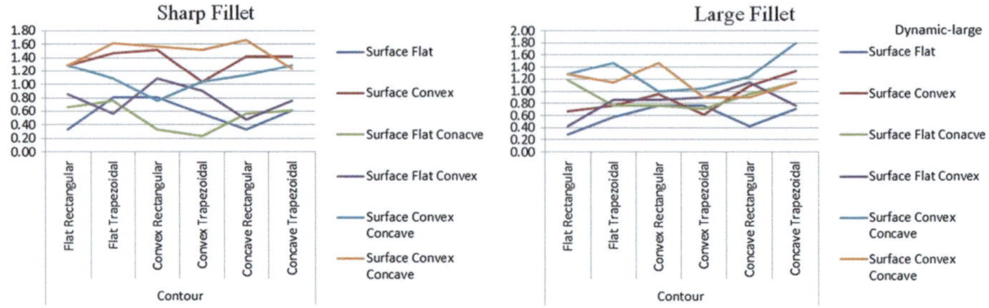

Fig. 9.43 Impact of visor shape on dynamism emotion

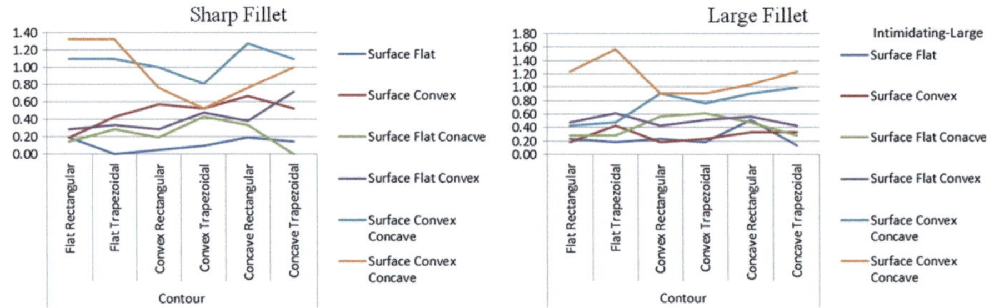

Fig. 9.44 Impact of visor shape on intimidation emotion

context as the intimidation score generated is low compared all other emotions. So in this limited context, convex-concave surface combination with flat contours and flat surfaces with concave contours are relatively more effective.

General trends from the above observation indicate the following:

a. Flat concave surfaces with small fillets generate sober character.
b. Rectangular headlights with convex contours with convex or convex-concave surfaces generate glamorous character.
c. Convex-convex surfaces with small fillets generate dynamic characters.

Convex-convex surfaces with flat on concave contours generate intimidating character.

9.5 Summary

Having discussed these four parts and their impact on the emotional character of a bike, there is one question that comes to our mind—is that all? Is there nothing else to the form of a bike that affects the character of a bike? Does the hugeness of a multi-cylinder not affect the form of a bike? Is the pattern of the wheel rim not part of the bike form? You can have many more such questions. Definitely, this is not all. Definitely, the form of an engine also affects the emotional character. But that basically constitutes the context of the bike and context has been discussed separately. Similarly, other questions have also been addressed in other contexts like balance, dynamism, etc. Here, in this chapter we have discussed only those parts which have been specifically styled to create the emotional character of the bike and which have the potential to drastically change the character.

Another potentially disruptive question is about the study itself. The study has been conducted through the representative geometric forms. The fundamental doubt that arises in our minds is whether these representative forms are actually representative. To be honest, the question can neither be answered in affirmative nor in negative. In fact, there cannot be any set of representative data which can accurately capture all possibilities. Otherwise, it will not be called a representative but actual data. But then, the basic issue of credibility of results remains. The answer lies in the representative character of the data and this is where we have to see that all possible combinations have been represented. This representativeness will be verified in later chapter when we tally the results with the actual results.

Appendix 1: Fuel Tank Shapes and their Emotion Scores

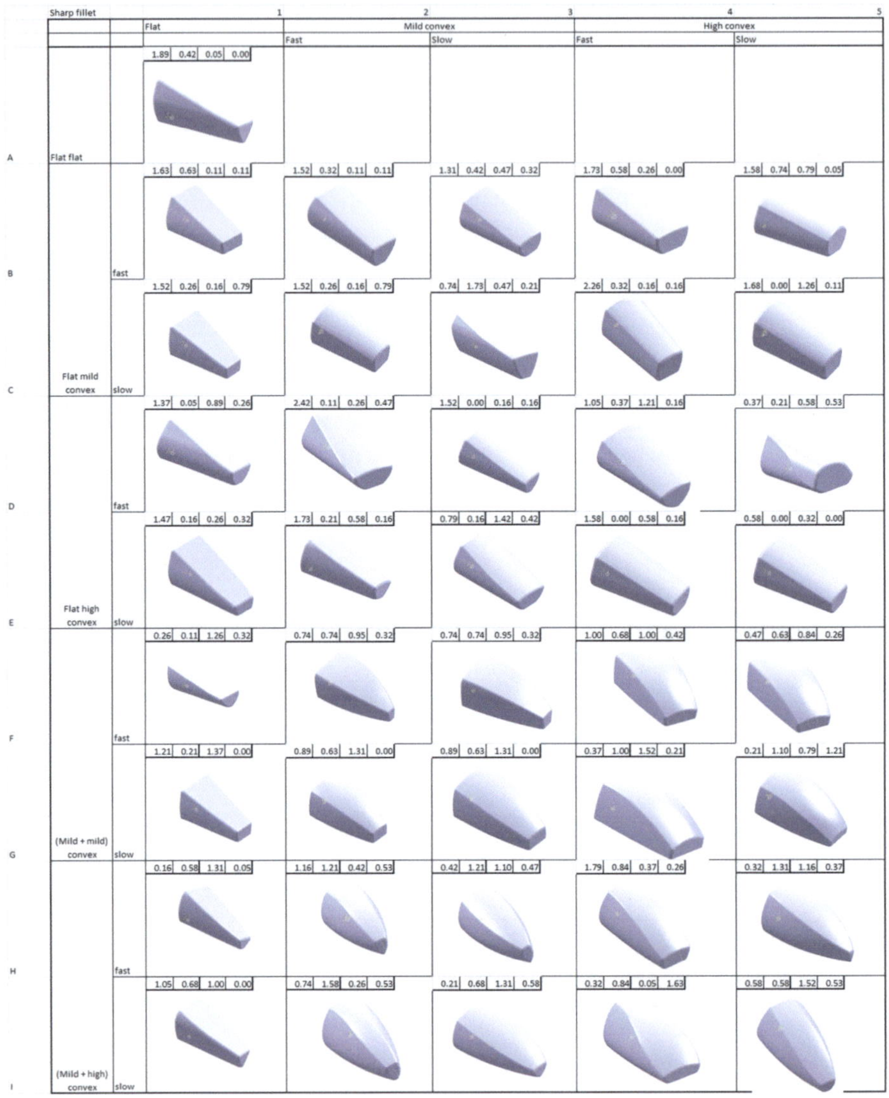

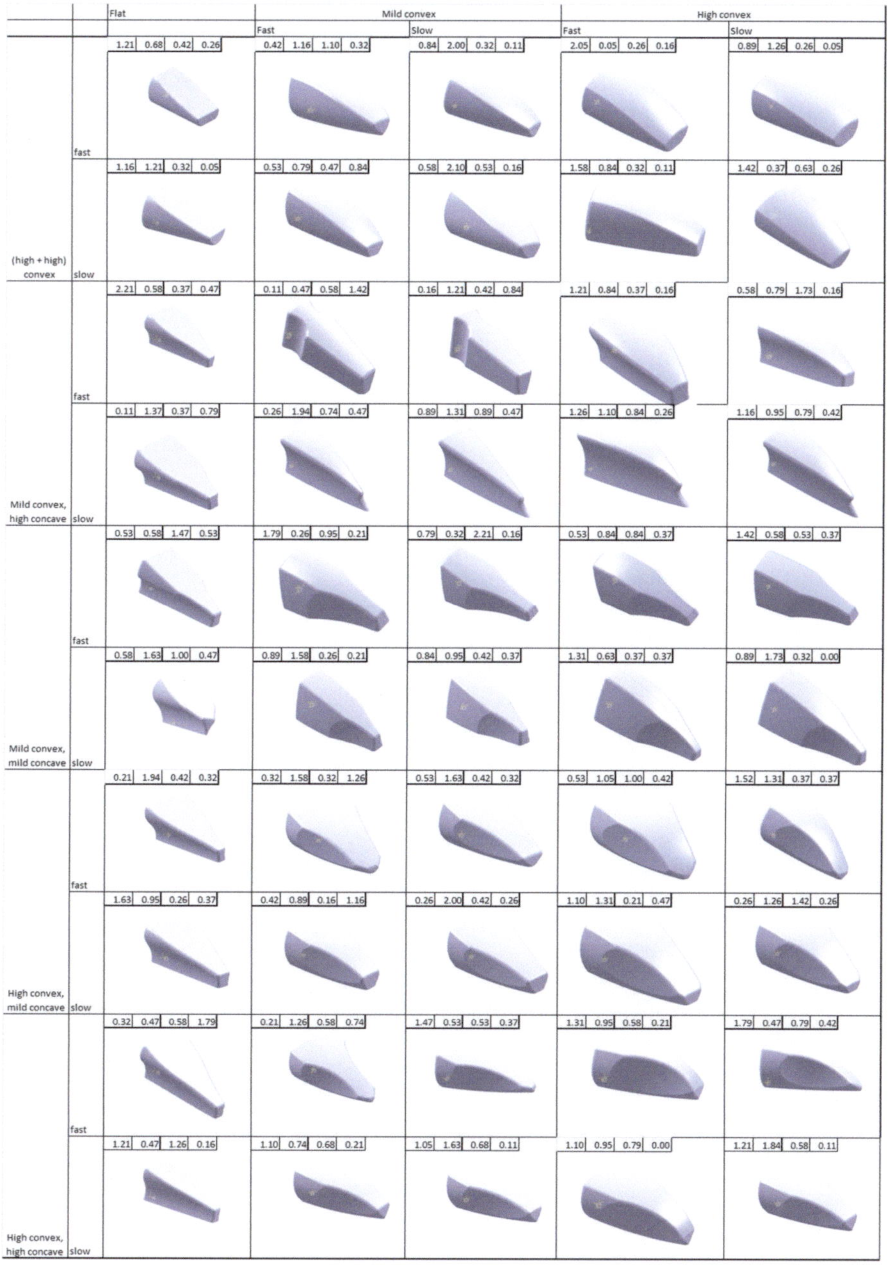

		Flat	Mild convex		High convex	
			Fast	Slow	Fast	Slow
(high + high) convex	fast	1.21 0.68 0.42 0.26	0.42 1.16 1.10 0.32	0.84 2.00 0.32 0.11	2.05 0.05 0.26 0.16	0.89 1.26 0.26 0.05
	slow	1.16 1.21 0.32 0.05	0.53 0.79 0.47 0.84	0.58 2.10 0.53 0.16	1.58 0.84 0.32 0.11	1.42 0.37 0.63 0.26
Mild convex, high concave	fast	2.21 0.58 0.37 0.47	0.11 0.47 0.58 1.42	0.16 1.21 0.42 0.84	1.21 0.84 0.37 0.16	0.58 0.79 1.73 0.16
	slow	0.11 1.37 0.37 0.79	0.26 1.94 0.74 0.47	0.89 1.31 0.89 0.47	1.26 1.10 0.84 0.26	1.16 0.95 0.79 0.42
Mild convex, mild concave	fast	0.53 0.58 1.47 0.53	1.79 0.26 0.95 0.21	0.79 0.32 2.21 0.16	0.53 0.84 0.84 0.37	1.42 0.58 0.53 0.37
	slow	0.58 1.63 1.00 0.47	0.89 1.58 0.26 0.21	0.84 0.95 0.42 0.37	1.31 0.63 0.37 0.37	0.89 1.73 0.32 0.00
High convex, mild concave	fast	0.21 1.94 0.42 0.32	0.32 1.58 0.32 1.26	0.53 1.63 0.42 0.32	0.53 1.05 1.00 0.42	1.52 1.31 0.37 0.37
	slow	1.63 0.95 0.26 0.37	0.42 0.89 0.16 1.16	0.26 2.00 0.42 0.26	1.10 1.31 0.21 0.47	0.26 1.26 1.42 0.26
High convex, high concave	fast	0.32 0.47 0.58 1.79	0.21 1.26 0.58 0.74	1.47 0.53 0.53 0.37	1.31 0.95 0.58 0.21	1.79 0.47 0.79 0.42
	slow	1.21 0.47 1.26 0.16	1.10 0.74 0.68 0.21	1.05 1.63 0.68 0.11	1.10 0.95 0.79 0.00	1.21 1.84 0.58 0.11

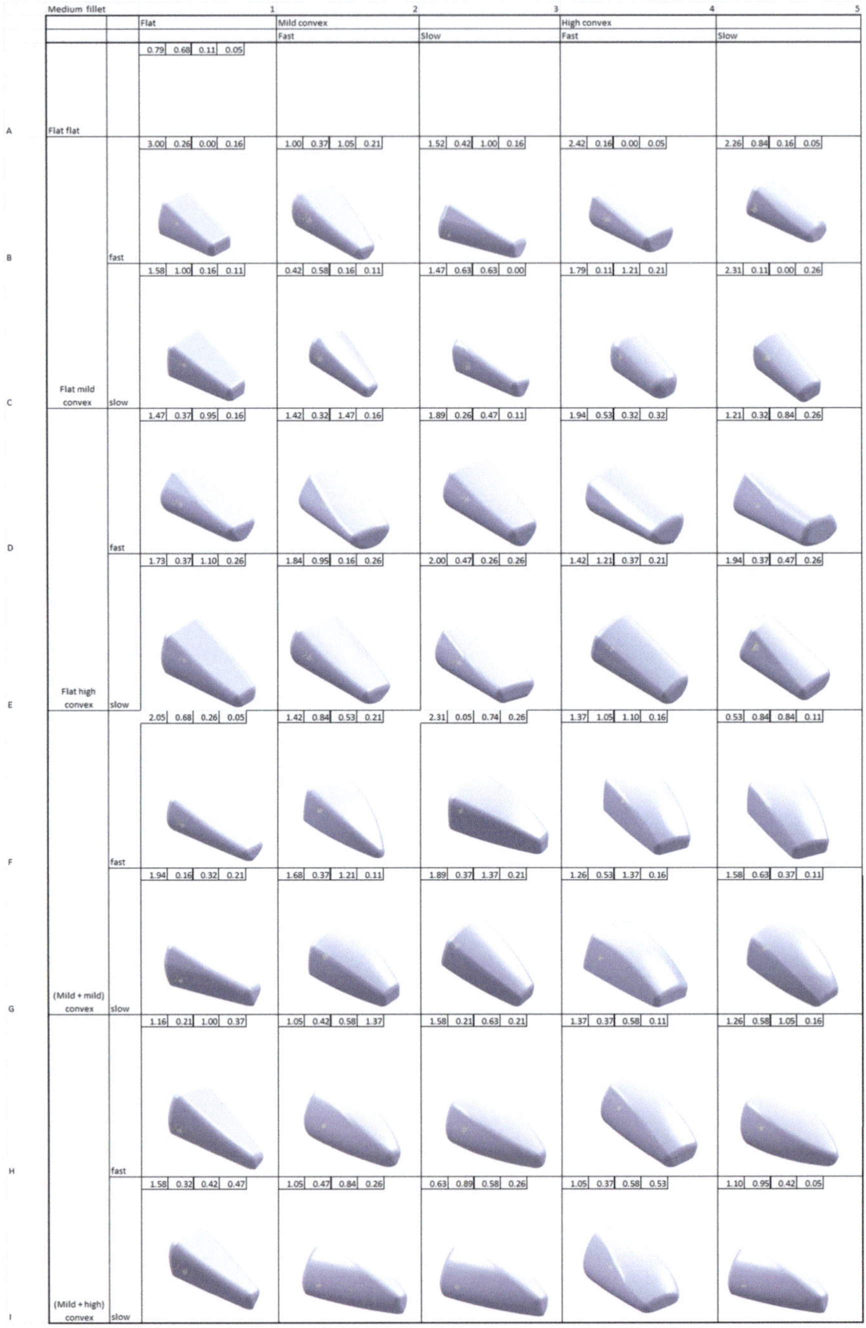
Medium fillet
1
2
3
4
5
Flat
Mild convex
Fast
Slow
High convex
Fast
Slow
A
Flat flat
0.79 0.68 0.11 0.05
B
fast
3.00 0.26 0.00 0.16
1.00 0.37 1.05 0.21
1.52 0.42 1.00 0.16
2.42 0.16 0.00 0.05
2.26 0.84 0.16 0.05
C
Flat mild convex
slow
1.58 1.00 0.16 0.11
0.42 0.58 0.16 0.11
1.47 0.63 0.63 0.00
1.79 0.11 1.21 0.21
2.31 0.11 0.00 0.26
D
fast
1.47 0.37 0.95 0.16
1.42 0.32 1.47 0.16
1.89 0.26 0.47 0.11
1.94 0.53 0.32 0.32
1.21 0.32 0.84 0.26
E
Flat high convex
slow
1.73 0.37 1.10 0.26
1.84 0.95 0.16 0.26
2.00 0.47 0.26 0.26
1.42 1.21 0.37 0.21
1.94 0.37 0.47 0.26
F
fast
2.05 0.68 0.26 0.05
1.42 0.84 0.53 0.21
2.31 0.05 0.74 0.26
1.37 1.05 1.10 0.16
0.53 0.84 0.84 0.11
G
(Mild + mild) convex
slow
1.94 0.16 0.32 0.21
1.68 0.37 1.21 0.11
1.89 0.37 1.37 0.21
1.26 0.53 1.37 0.16
1.58 0.63 0.37 0.11
H
fast
1.16 0.21 1.00 0.37
1.05 0.42 0.58 1.37
1.58 0.21 0.63 0.21
1.37 0.37 0.58 0.11
1.26 0.58 1.05 0.16
I
(Mild + high) convex
slow
1.58 0.32 0.42 0.47
1.05 0.47 0.84 0.26
0.63 0.89 0.58 0.26
1.05 0.37 0.58 0.53
1.10 0.95 0.42 0.05

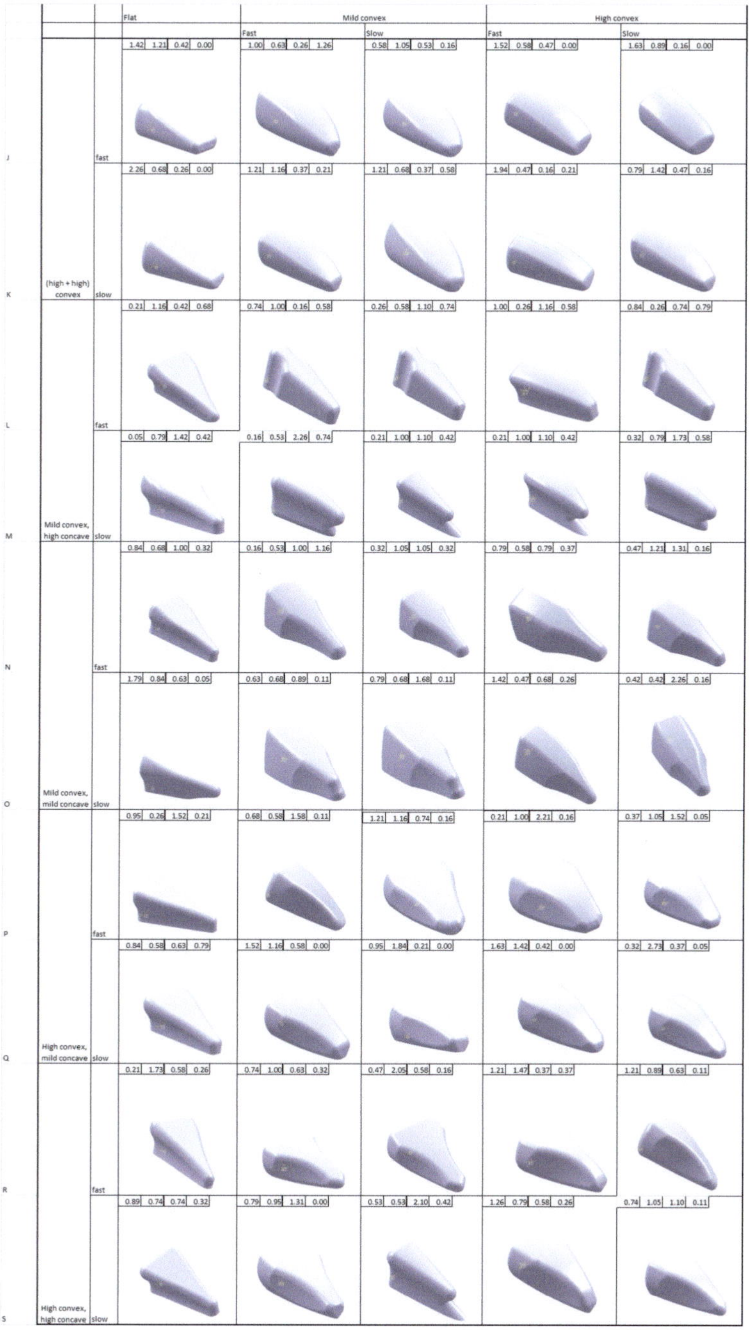

			Flat	Mild convex Fast	Mild convex Slow	High convex Fast	High convex Slow
J	(high + high) convex	fast	1.42 1.21 0.42 0.00	1.00 0.63 0.26 1.26	0.58 1.05 0.53 0.16	1.52 0.58 0.47 0.00	1.63 0.89 0.16 0.00
K		slow	2.26 0.68 0.26 0.00	1.21 1.16 0.37 0.21	1.21 0.68 0.37 0.58	1.94 0.47 0.16 0.21	0.79 1.42 0.47 0.16
L	Mild convex, high concave	fast	0.21 1.16 0.42 0.68	0.74 1.00 0.16 0.58	0.26 0.58 1.10 0.74	1.00 0.26 1.16 0.58	0.84 0.26 0.74 0.79
M		slow	0.05 0.79 1.42 0.42	0.16 0.53 2.26 0.74	0.21 1.00 1.10 0.42	0.21 1.00 1.10 0.42	0.32 0.79 1.73 0.58
N	Mild convex, mild concave	fast	0.84 0.68 1.00 0.32	0.16 0.53 1.00 1.16	0.32 1.05 1.05 0.32	0.79 0.58 0.79 0.37	0.47 1.21 1.31 0.16
O		slow	1.79 0.84 0.63 0.05	0.63 0.68 0.89 0.11	0.79 0.68 1.68 0.11	1.42 0.47 0.68 0.26	0.42 0.42 2.26 0.16
P	High convex, mild concave	fast	0.95 0.26 1.52 0.21	0.68 0.58 1.58 0.11	1.21 1.16 0.74 0.16	0.21 1.00 2.21 0.16	0.37 1.05 1.52 0.05
Q		slow	0.84 0.58 0.63 0.79	1.52 1.16 0.58 0.00	0.95 1.84 0.21 0.00	1.63 1.42 0.42 0.00	0.32 2.73 0.37 0.05
R	High convex, high concave	fast	0.21 1.73 0.58 0.26	0.74 1.00 0.63 0.32	0.47 2.05 0.58 0.16	1.21 1.47 0.37 0.37	1.21 0.89 0.63 0.11
S		slow	0.89 0.74 0.74 0.32	0.79 0.95 1.31 0.00	0.53 0.53 2.10 0.42	1.26 0.79 0.58 0.26	0.74 1.05 1.10 0.11

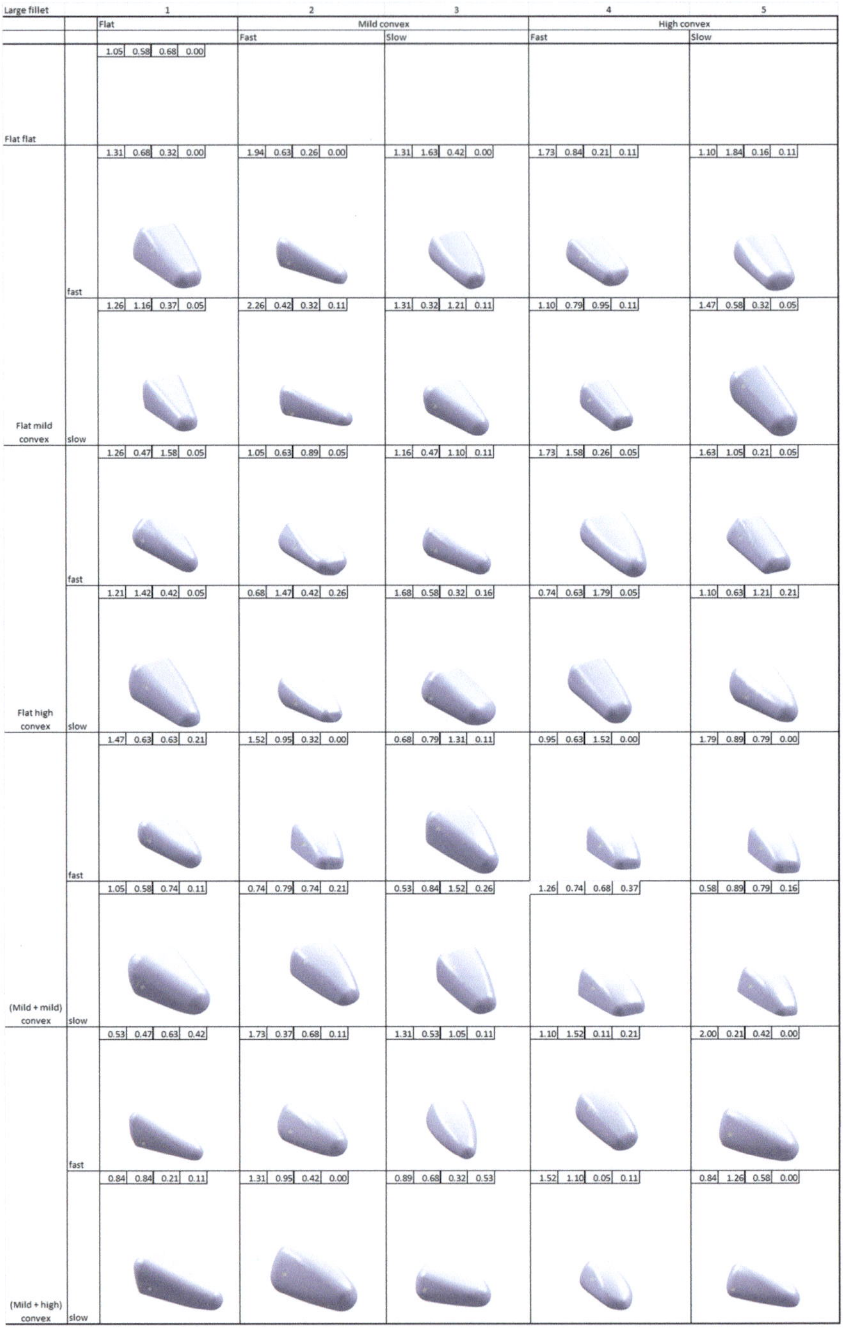
Large fillet
1
2
3
4
5
Flat
Mild convex
High convex
Fast
Slow
Fast
Slow
1.05 0.58 0.68 0.00
Flat flat
1.31 0.68 0.32 0.00
1.94 0.63 0.26 0.00
1.31 1.63 0.42 0.00
1.73 0.84 0.21 0.11
1.10 1.84 0.16 0.11
fast
1.26 1.16 0.37 0.05
2.26 0.42 0.32 0.11
1.31 0.32 1.21 0.11
1.10 0.79 0.95 0.11
1.47 0.58 0.32 0.05
Flat mild convex
slow
1.26 0.47 1.58 0.05
1.05 0.63 0.89 0.05
1.16 0.47 1.10 0.11
1.73 1.58 0.26 0.05
1.63 1.05 0.21 0.05
fast
1.21 1.42 0.42 0.05
0.68 1.47 0.42 0.26
1.68 0.58 0.32 0.16
0.74 0.63 1.79 0.05
1.10 0.63 1.21 0.21
Flat high convex
slow
1.47 0.63 0.63 0.21
1.52 0.95 0.32 0.00
0.68 0.79 1.31 0.11
0.95 0.63 1.52 0.00
1.79 0.89 0.79 0.00
fast
1.05 0.58 0.74 0.11
0.74 0.79 0.74 0.21
0.53 0.84 1.52 0.26
1.26 0.74 0.68 0.37
0.58 0.89 0.79 0.16
(Mild + mild) convex
slow
0.53 0.47 0.63 0.42
1.73 0.37 0.68 0.11
1.31 0.53 1.05 0.11
1.10 1.52 0.11 0.21
2.00 0.21 0.42 0.00
fast
0.84 0.84 0.21 0.11
1.31 0.95 0.42 0.00
0.89 0.68 0.32 0.53
1.52 1.10 0.05 0.11
0.84 1.26 0.58 0.00
(Mild + high) convex
slow

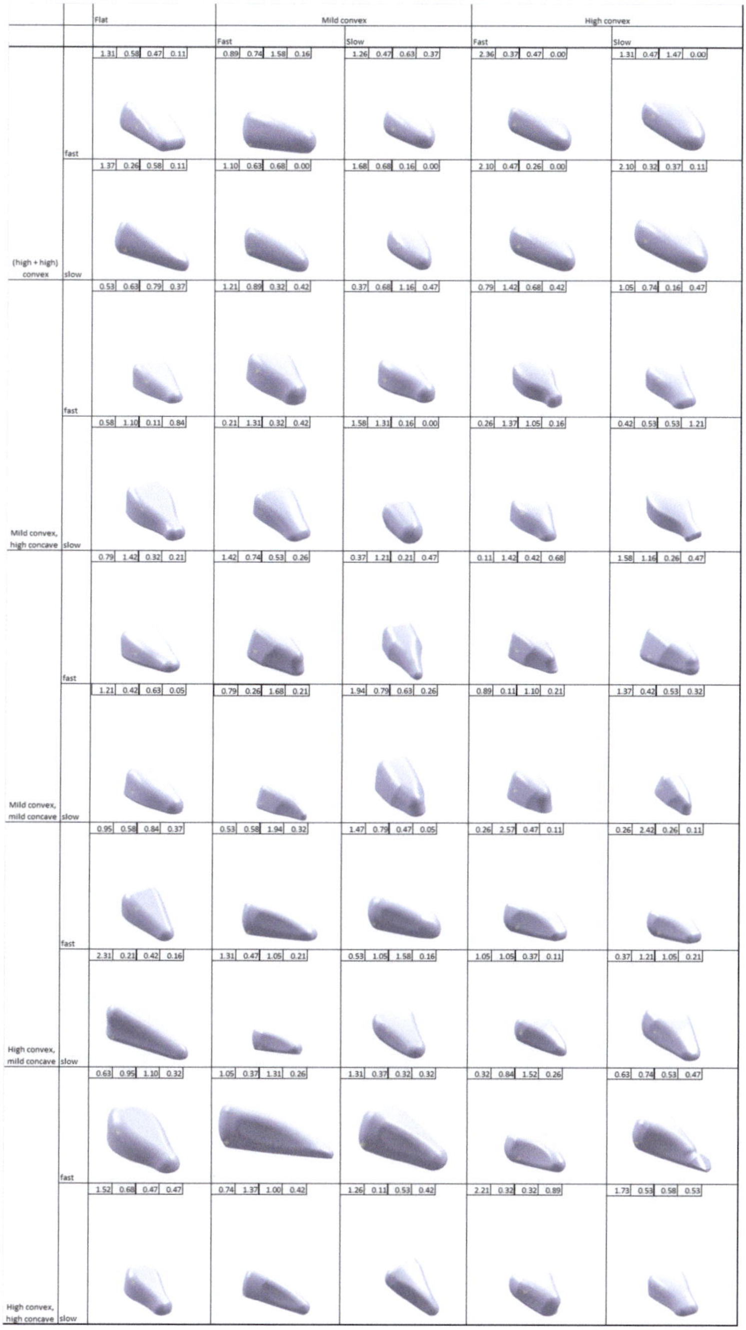

		Flat	Mild convex		High convex	
			Fast	Slow	Fast	Slow
(high + high) convex	fast	1.31 0.58 0.47 0.11	0.89 0.74 1.58 0.16	1.26 0.47 0.63 0.37	2.36 0.37 0.47 0.00	1.31 0.47 1.47 0.00
	slow	1.37 0.26 0.58 0.11	1.10 0.63 0.68 0.00	1.68 0.68 0.16 0.00	2.10 0.47 0.26 0.00	2.10 0.32 0.37 0.11
Mild convex, high concave	fast	0.53 0.63 0.79 0.37	1.21 0.89 0.32 0.42	0.37 0.68 1.16 0.47	0.79 1.42 0.68 0.42	1.05 0.74 0.16 0.47
	slow	0.58 1.10 0.11 0.84	0.21 1.31 0.32 0.42	1.58 1.31 0.16 0.00	0.26 1.37 1.05 0.16	0.42 0.53 0.53 1.21
Mild convex, mild concave	fast	0.79 1.42 0.32 0.21	1.42 0.74 0.53 0.26	0.37 1.21 0.21 0.47	0.11 1.42 0.42 0.68	1.58 1.16 0.26 0.47
	slow	1.21 0.42 0.63 0.05	0.79 0.26 1.68 0.21	1.94 0.79 0.63 0.26	0.89 0.11 1.10 0.21	1.37 0.42 0.53 0.32
High convex, mild concave	fast	0.95 0.58 0.84 0.37	0.53 0.58 1.94 0.32	1.47 0.79 0.47 0.05	0.26 2.57 0.47 0.11	0.26 2.42 0.26 0.11
	slow	2.31 0.21 0.42 0.16	1.31 0.47 1.05 0.21	0.53 1.05 1.58 0.16	1.05 1.05 0.37 0.11	0.37 1.21 1.05 0.21
High convex, high concave	fast	0.63 0.95 1.10 0.32	1.05 0.37 1.31 0.26	1.31 0.37 0.32 0.32	0.32 0.84 1.52 0.26	0.63 0.74 0.53 0.47
	slow	1.52 0.68 0.47 0.47	0.74 1.37 1.00 0.42	1.26 0.11 0.53 0.42	2.21 0.32 0.32 0.89	1.73 0.53 0.58 0.53

Appendix 2: Side Cover Shapes and their Emotion Scores

SIDE COVER	Triangular								Polygon +/+								Polygon +/-							
	sharp fillet				large fillet				sharp fillet				large fillet				sharp fillet				large fillet			
	S	G	D	I	S	G	D	I	S	G	D	I	S	G	D	I	S	G	D	I	S	G	D	I
Flat 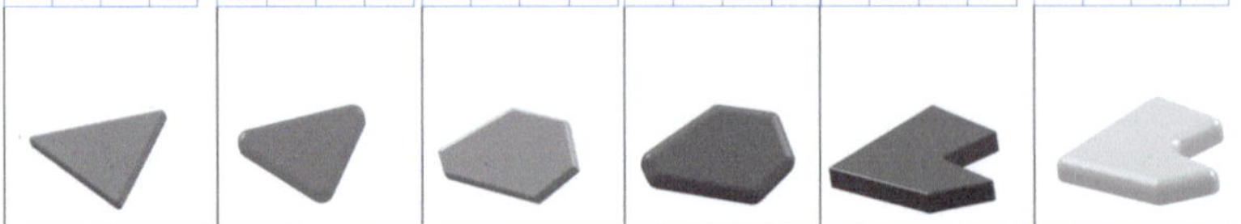	2.65	0.00	0.49	0.05	3.00	0.39	0.39	0.25	2.11	0.74	0.64	0.15	1.47	1.86	0.88	0.15	1.52	1.18	1.18	0.29	1.32	1.57	1.03	0.20
Convex 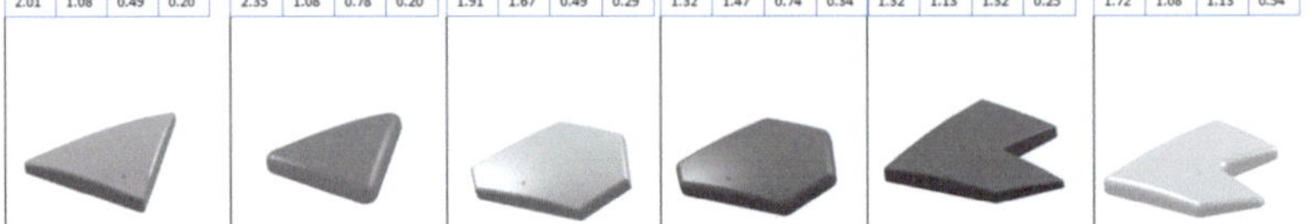	2.01	1.08	0.49	0.20	2.35	1.08	0.78	0.20	1.91	1.67	0.49	0.29	1.32	1.47	0.74	0.34	1.32	1.13	1.32	0.25	1.72	1.08	1.13	0.54
Flat / Flat-angle 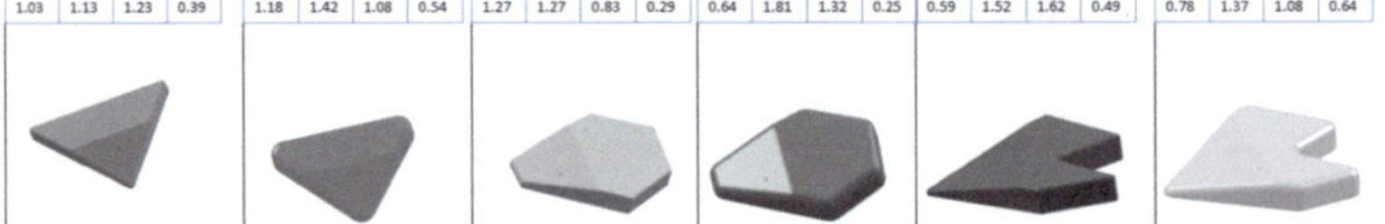	1.03	1.13	1.23	0.39	1.18	1.42	1.08	0.54	1.27	1.27	0.83	0.29	0.64	1.81	1.32	0.25	0.59	1.52	1.62	0.49	0.78	1.37	1.08	0.64
Flat / Flat-step 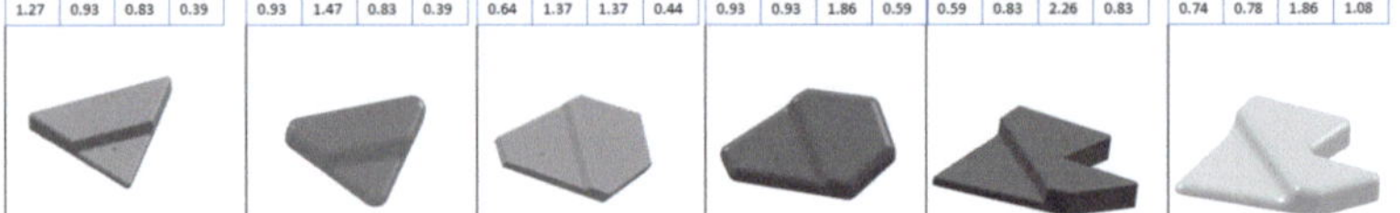	1.27	0.93	0.83	0.39	0.93	1.47	0.83	0.39	0.64	1.37	1.37	0.44	0.93	0.93	1.86	0.59	0.59	0.83	2.26	0.83	0.74	0.78	1.86	1.08
Flat concave 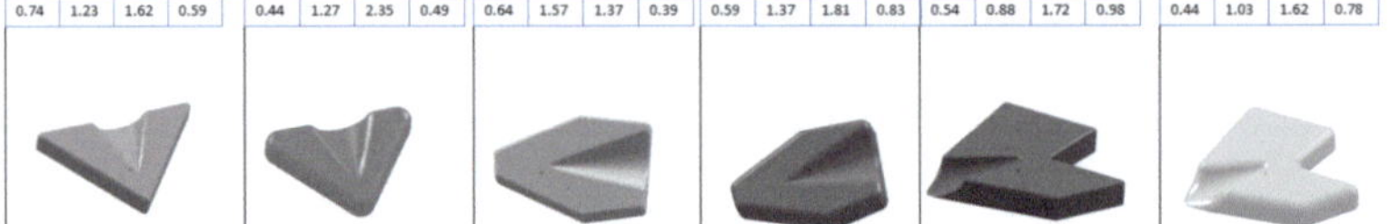	0.74	1.23	1.62	0.59	0.44	1.27	2.35	0.49	0.64	1.57	1.37	0.39	0.59	1.37	1.81	0.83	0.54	0.88	1.72	0.98	0.44	1.03	1.62	0.78
Flat convex 	0.74	0.74	2.11	0.25	0.88	1.13	1.47	0.69	0.78	0.54	2.06	0.78	0.54	0.88	2.06	0.69	0.44	1.03	1.57	1.23	0.20	1.27	1.81	0.88
Convex-concave 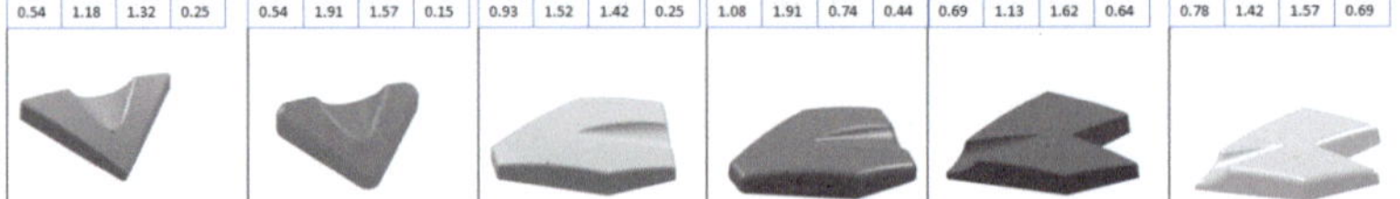	0.54	1.18	1.32	0.25	0.54	1.91	1.57	0.15	0.93	1.52	1.42	0.25	1.08	1.91	0.74	0.44	0.69	1.13	1.62	0.64	0.78	1.42	1.57	0.69

SIDE COVER	Recto-curvilinear +/+								Recto-curvilinear +/-							
	sharp fillet				large fillet				sharp fillet				large fillet			
	S	G	D	I	S	G	D	I	S	G	D	I	S	G	D	I
Flat	2.65	1.03	0.34	0.15	0.09	0.74	2.41	1.25	0.09	0.74	2.41	1.25	0.09	0.74	2.41	1.25
													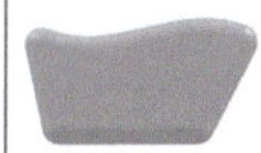			
Convex	1.96	0.74	0.69	0.29	1.96	0.98	0.69	0.44	1.77	1.23	0.74	0.25	1.47	1.47	0.78	0.15

Flat / Flat-angle	1.08	1.08	1.27	0.69	1.27	1.47	1.03	0.49	1.57	1.13	0.69	0.44	1.42	1.27	0.69	0.29

Flat / Flat-step	1.47	0.59	0.93	0.69	1.47	1.27	0.83	0.54	1.52	1.27	0.98	0.34	2.01	1.57	0.54	0.44
													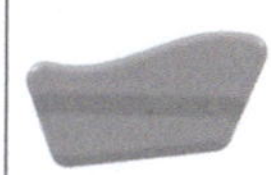			
Flat concave	1.18	1.32	0.88	0.15	1.08	1.81	0.93	0.20	1.47	1.62	0.64	0.25	1.42	1.52	0.93	0.25
													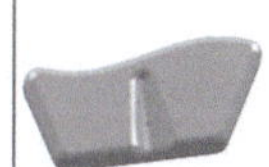			
Flat convex	0.83	0.88	1.91	0.44	1.67	1.13	0.88	0.64	1.08	1.37	0.64	0.74	0.69	1.77	0.69	0.64
					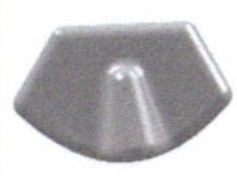								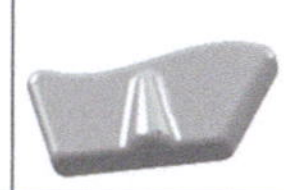			
Convex-concave	0.98	1.23	1.13	0.49	1.18	1.57	0.83	0.39	1.13	1.86	0.59	0.44	1.08	1.42	0.64	0.39
													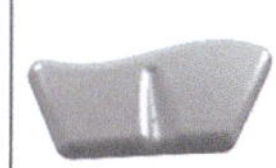			

Appendix 3: Seat Shapes and their Emotion Scores

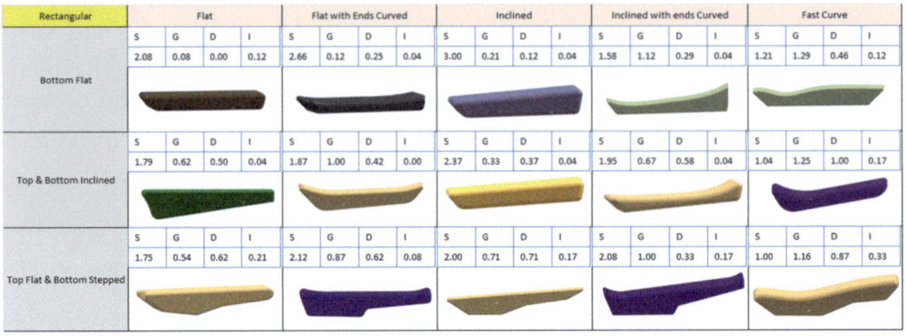

Rectangular	Flat				Flat with Ends Curved				Inclined				Inclined with ends Curved				Fast Curve			
	S	G	D	I	S	G	D	I	S	G	D	I	S	G	D	I	S	G	D	I
Bottom Flat	2.08	0.08	0.00	0.12	2.66	0.12	0.25	0.04	3.00	0.21	0.12	0.04	1.58	1.12	0.29	0.04	1.21	1.29	0.46	0.12
Top & Bottom Inclined	1.79	0.62	0.50	0.04	1.87	1.00	0.42	0.00	2.37	0.33	0.37	0.04	1.95	0.67	0.58	0.04	1.04	1.25	1.00	0.17
Top Flat & Bottom Stepped	1.75	0.54	0.62	0.21	2.12	0.87	0.62	0.08	2.00	0.71	0.71	0.17	2.08	1.00	0.33	0.17	1.00	1.16	0.87	0.33

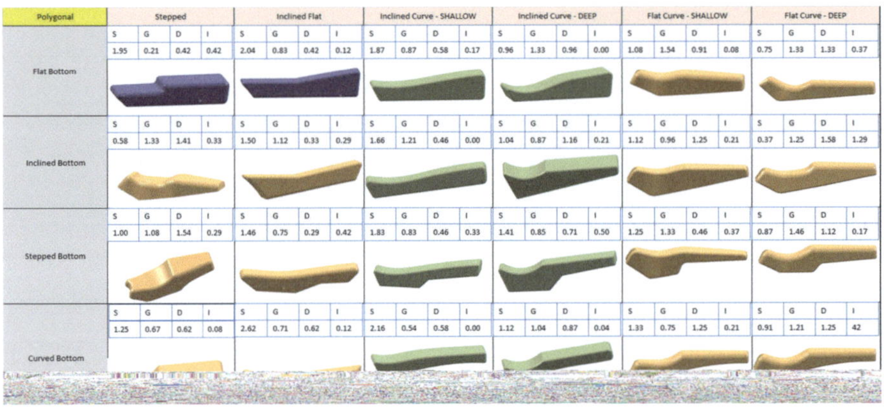

Polygonal	Stepped				Inclined Flat				Inclined Curve - SHALLOW				Inclined Curve - DEEP				Flat Curve - SHALLOW				Flat Curve - DEEP			
	S	G	D	I	S	G	D	I	S	G	D	I	S	G	D	I	S	G	D	I	S	G	D	I
Flat Bottom	1.95	0.21	0.42	0.42	2.04	0.83	0.42	0.12	1.87	0.87	0.58	0.17	0.96	1.33	0.96	0.00	1.08	1.54	0.91	0.08	0.75	1.33	1.33	0.37
Inclined Bottom	0.58	1.33	1.41	0.33	1.50	1.12	0.33	0.29	1.66	1.21	0.46	0.00	1.04	0.87	1.16	0.21	1.12	0.96	1.25	0.21	0.37	1.25	1.58	1.29
Stepped Bottom	1.00	1.08	1.54	0.29	1.46	0.75	0.29	0.42	1.83	0.83	0.46	0.33	1.41	0.85	0.71	0.50	1.25	1.33	0.46	0.37	0.87	1.46	1.12	0.17
Curved Bottom	1.25	0.67	0.62	0.08	2.62	0.71	0.62	0.12	2.16	0.54	0.58	0.00	1.12	1.04	0.87	0.04	1.33	0.75	1.25	0.21	0.91	1.21	1.25	42

Appendix 4: Visor Shapes and their Emotion Scores

VISOR (small fillet)	FLAT CONTOUR								CONVEX CONTOUR								CONCAVE CONTOUR							
	RECTNGULAR H/L				TRAPEZOIDAL H/L				RECTNGULAR H/L				TRAPEZOIDAL H/L				RECTNGULAR H/L				TRAPEZOIDAL H/L			
	S	G	D	I	S	G	D	I	S	G	D	I	S	G	D	I	S	G	D	I	S	G	D	I
FLAT	2.66	0.19	0.33	0.19	2.51	0.33	0.81	0.00	2.66	0.57	0.81	0.05	2.23	1.00	0.57	0.09	2.56	1.00	0.33	0.19	2.18	0.95	0.62	0.14
	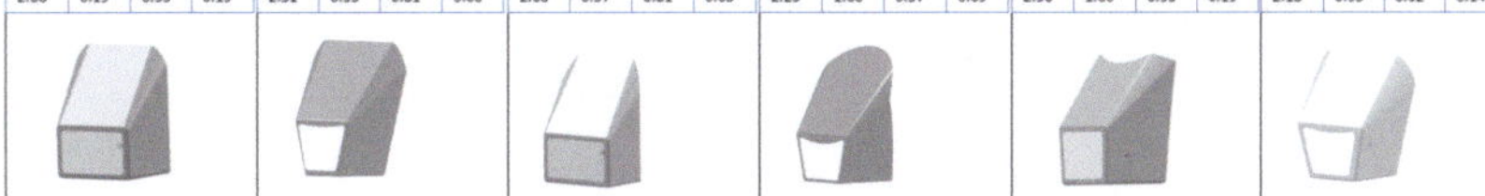																							
CONVEX	1.47	1.66	1.28	0.19	1.66	.23	1.47	0.43	0.81	1.76	1.52	0.57	1.19	1.47	1.04	0.52	1.42	0.95	1.42	0.66	1.04	1.66	1.42	0.52
	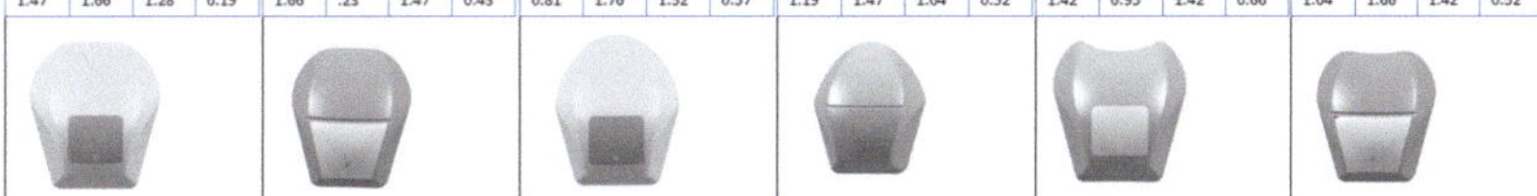																							
FLAT CONCAVE	2.89	0.76	0.66	0.14	2.61	0.47	0.76	0.28	2.94	0.52	0.33	0.19	3.00	0.90	0.24	0.43	2.42	0.62	0.57	0.33	3.00	0.66	0.62	0.00
	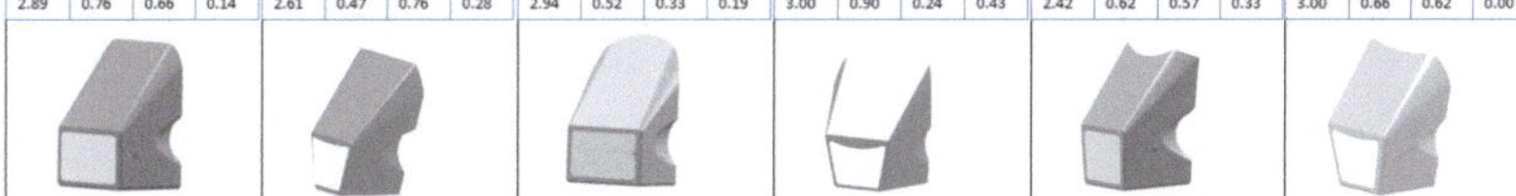																							
FLAT CONVEX	1.90	0.57	0.85	0.28	2.28	0.76	0.57	0.33	1.85	0.38	1.09	0.28	2.00	0.66	0.90	0.47	2.09	0.95	0.47	0.38	1.57	1.19	0.76	0.71
	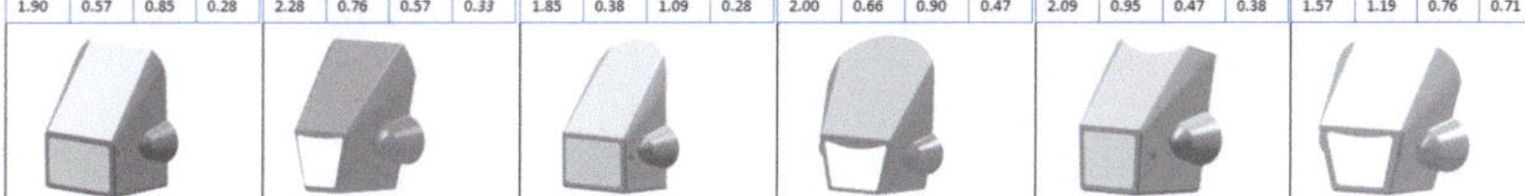																							
CONVEX CONCAE	0.71	1.47	1.28	1.09	1.00	1.71	1.09	1.09	0.85	1.80	0.76	1.00	0.76	1.47	1.04	0.81	0.52	1.71	1.14	1.28	0.47	1.66	1.28	1.09
	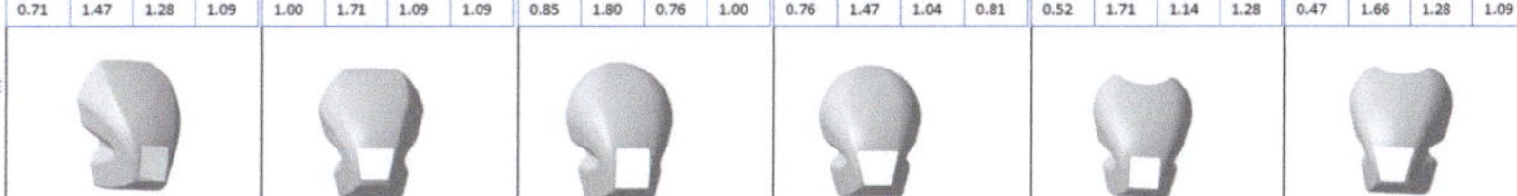																							
CONVEX CONVEX	0.52	1.28	1.28	1.33	0.57	1.33	1.61	1.33	1.14	1.04	1.57	0.76	0.81	1.28	1.52	52	0.47	1.57	1.66	0.76	1.00	1.28	1.23	1.00
	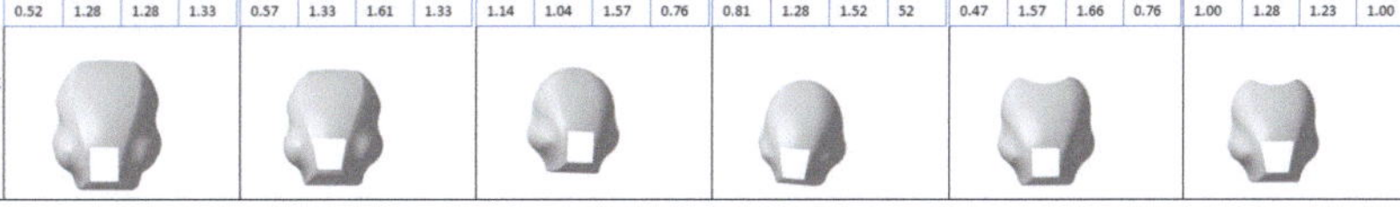																							

VISOR (small fillet)	FLAT CONTOUR								CONVEX CONTOUR								CONCAVE CONTOUR							
	RECTNGULAR H/L				TRAPEZOIDAL H/L				RECTNGULAR H/L				TRAPEZOIDAL H/L				RECTNGULAR H/L				TRAPEZOIDAL H/L			
	S	G	D	I	S	G	D	I	S	G	D	I	S	G	D	I	S	G	D	I	S	G	D	I
FLAT	2.89	0.43	0.28	0.24	2.75	0.57	0.57	0.19	2.32	0.52	0.76	0.24	2.42	1.00	0.76	0.19	2.37	0.28	0.43	0.52	1.99	0.95	0.71	0.14

CONVEX	1.80	1.52	0.66	0.19	1.23	1.57	0.76	0.43	1.71	1.80	0.95	0.19	1.66	1.71	0.62	0.24	1.71	1.04	1.09	0.33	1.57	1.09	1.33	0.33
	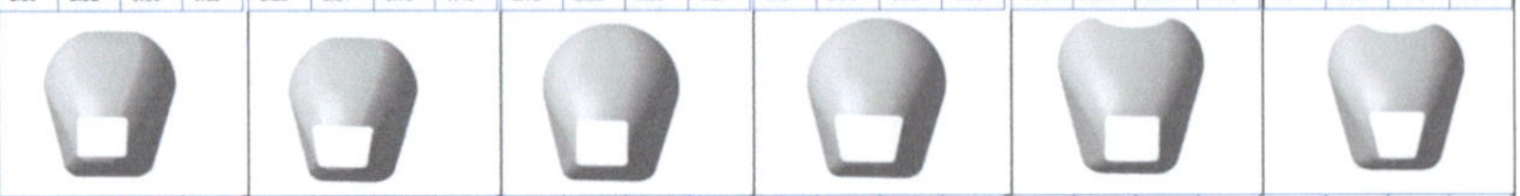																							
FLAT CONCAVE	1.38	1.23	1.19	0.28	2.04	0.90	0.76	0.28	2.09	0.90	0.76	0.57	2.09	1.00	0.71	0.62	2.28	0.47	0.95	0.47	1.52	0.85	1.14	0.28
	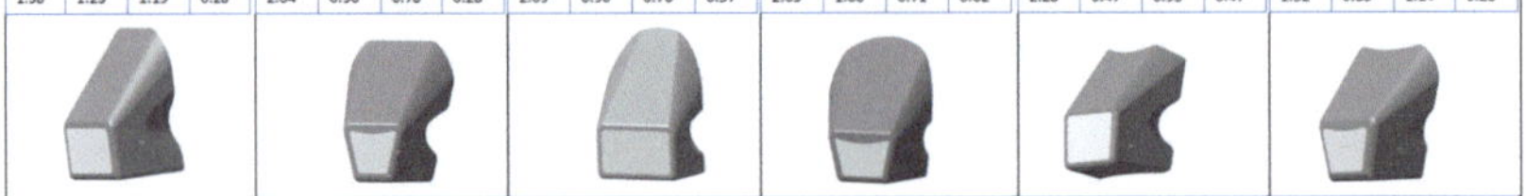																							
FLAT CONVEX	2.18	0.81	0.43	0.47	2.04	0.85	0.85	0.62	1.94	0.85	0.85	0.43	1.47	1.42	0.90	0.52	1.61	1.14	0.90	0.57	1.71	1.23	0.76	0.43
	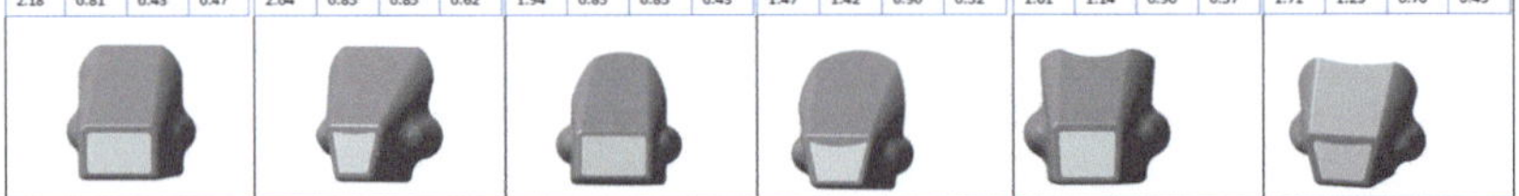																							
CONVEX CONCAE	1.19	1.42	1.28	0.43	1.04	1.09	1.47	0.47	1.04	1.19	1.00	0.90	1.04	1.28	1.04	0.76	0.66	1.52	1.23	0.90	0.43	1.42	1.80	1.00
	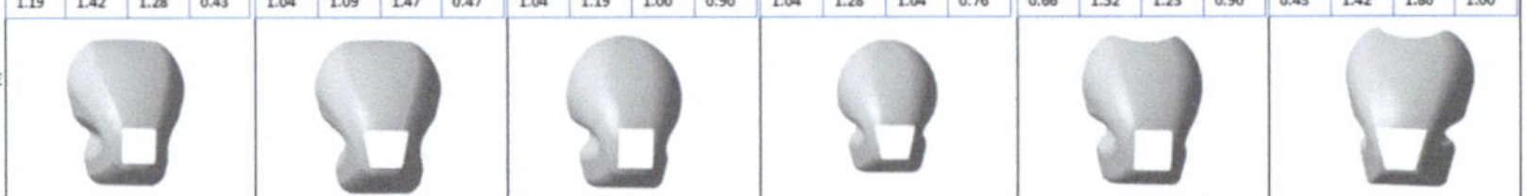																							
CONVEX CONVEX	0.90	0.81	1.28	1.23	1.04	0.76	1.14	1.57	0.47	1.47	1.47	0.90	0.85	1.47	0.90	0.90	1.38	0.95	0.90	1.04	0.90	1.14	1.14	1.23
	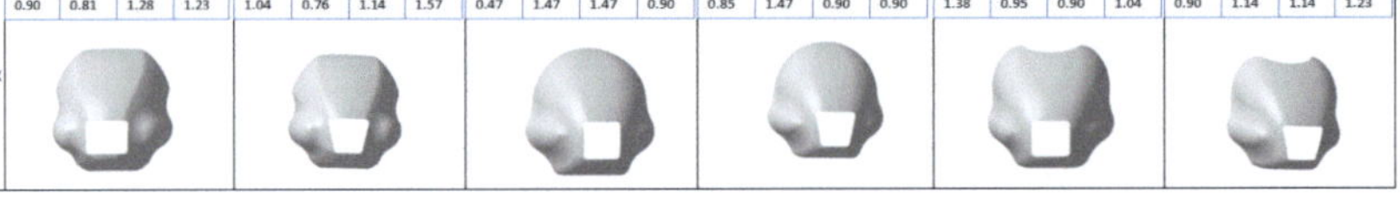																							

Chapter 10
Color and Texture

Should your glance on mornings lovely
Lift to drink the heaven's blue
Or when sun, veiled by sirocco,
Royal red sinks out of view –
Give to Nature praise and honor.
Blithe of heart and sound of eye,
Knowing for the world of color
Where its broad foundations lie.

—Goethe

Goethe, the famous poet came out with his science of colors in 1810. This theory can be largely divided into two parts. The first part challenges the Newtonian theory of light breaking into a spectrum of colors while passing through a prism due to each color having a different refractive index. The second part, which is of more interest to us, deals with the psychology of colors (see Fig. 10.1 and Table 10.1). Though this was later ridiculed by many for not having any scientific basis, it is an important event as, for the first time, someone tried to link colors to emotions on an objective basis.

The fundamental question we need to answer at this point is—why do we need to reinvent the wheel when we already have the template in form of wheel of colors by Goethe. The first thing is, as many had pointed out in his lifetime only, is that it has no scientific basis. The second and more important factor is that the impact of colors, or for that matter any design factor on emotion is driven by civilizational memories. So, it cannot be same for all societies and has to be separately derived for each society. This is what we will do in this chapter.

We have discussed the complexities of color and texture in earlier chapters. But these complexities multiply when we get to study the impact of color on emotion. Just look at the numbers. There are thousands of hues. Each hue has 12 levels of chroma and 10 levels of value. Then we have the combination of colors. Even if we take a combination of two colors, we get more than a million of combinations. But in real life, rarely do we get a two-color combination. Most often, it is more than three. So we have a situation where we dive into an ocean to collect a needle and we

S. Chandra, *Aesthetics: Quantification and Deconstruction*,
Design Science and Innovation, https://doi.org/10.1007/978-981-10-6235-3_10

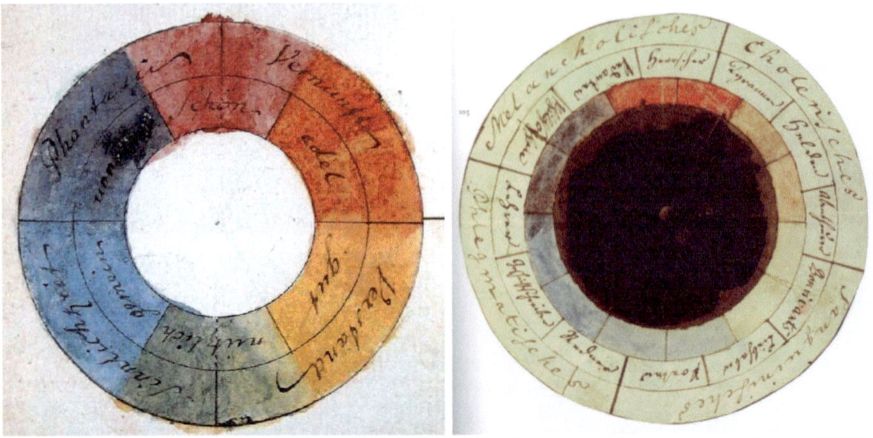

The left figure shows Goethe's wheel of color showing the emotions attached to the colors. The words are in German and you can see the details in the table shown below. The right figure shows the "rose of temperaments" (*Temperamentenrose*), an earlier study (1798/9) by Goethe and Schiller, matching twelve colours to human occupations or their character traits (*tyrants, heroes, adventurers, hedonists, lovers, poets, public speakers, historians, teachers, philosophers, pedants, rulers*), grouped in the four temperament

Fig. 10.1 Goethe's color theory in original

Table 10.1 Translations for Fig. 10.1

German	English	Symbolism
Purpur	Magenta (or purple)	Schon (beautiful)
Rot	Red	
Gelbrot	Orange	Edel (noble)
Orange		
Gulb	Yellow	Gut (good)
Grun	Green	Nutzlich (useful)
Blau	Blue	Gemein (mean, common)
Violett	Violet	Unnothig (unnecessary)
Blaurot		

find that one life is not enough to search through the ocean. So, the challenge before us is to limit the ocean into a pond and still keep the diversity of the ocean.

First, we consider the hues. For our purpose, we limit to the primary colors—red, blue, green, and yellow. We have to add black, white and gray—it is elementary. If we have three levels of value (high, medium and low) and similar three levels of chroma, still we have nine combinations for each hue, which leaves us with 2262 combination, even if we limit the number of colors to two in each combination. This is not only non-manageable but non-inferable as well. So we limit to total three combinations of value and chroma—high value high chroma, medium value

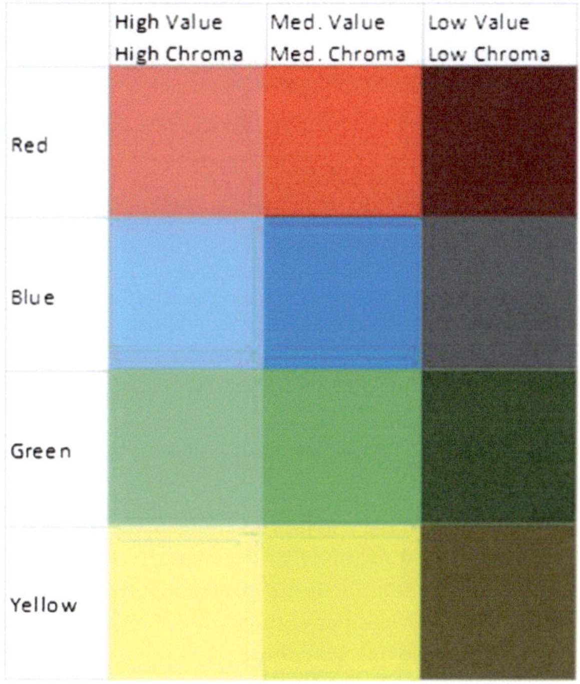

Fig. 10.2 Color palette for hue value and chroma

medium chroma and low value low chroma. This drastically simplifies our problem and leaves us with 331 combinations which are manageable and inferable.

As we go through the palette of colors (see Fig. 10.2), we can simply replace the long-winded labels like high value high chroma, etc., with light (for high value high chroma), deep (medium value medium chroma), and dark (low value low chroma) shades. It is clear that black and white have no degrees of lightness or darkness and we keep only the middle level of gray for sake of simplicity.

Since in this chapter, we will be discussing the impact of color on emotions through these chosen shades, it will be appropriate to briefly describe the methodology. The fundamental objects of this study are color plates. An interesting point of discussion can be, that, why do not we use motorcycle in various colors rather than using rather disinteresting plates of color. Answer is simple. Since, the objective is to de-couple the other aspects of motorcycle design from the impact of colors; images of motorcycles will defeat the purpose.

The basic methodology is simple. The plates are shown to respondents, who respond with the type of emotion generated and the intensity of emotion on a scale of 1–3. By collating the data, we come to SGDI score for each plate which represents the hue, value, and chroma combination for the color. Let us be very clear, the SGDI score are true only for the society, in which the study has been done. Any change in geography or the social milieu needs another survey.

Fig. 10.3 Sample color plate for *deep green*

Coming back to plates, three types of plates are created. The first group of plates is that of the primary colors in light, deep, and dark shades plus white, gray, and black colors. An example of the plates is shown in Fig. 10.3.

Now, let us talk about the nomenclature of colors. You find, on left side bottom of the plate, the number signifying the serial number of the plate and on the right side bottom the nomenclature signifying the color and the shade. Needles to explain, the letter preceding the number represents the color (g for green, b for blue, r for red and y for yellow and black, white and gray represented by bl, w, and gr) and the number represents the shade (1 for light, 2 for deep and 3 for dark).

10.1 Mono-Colors

For this group of plates, the SGDI score are plotted for sober, glamorous, dynamic, and intimidating emotions and they reveal the trends for their relationship with these four emotions. These trends are shown in Fig. 10.4.

Look at the "sober" curves and you will find them interesting. They are interesting because the primary colors at the top of the curves generate best of sobriety at their lightest and darkest, where as deep shades of these colors reduce its level somewhat. These colors happen to be those which fall somewhere in the middle of the spectrum—blue and yellow. Expectedly, white is the eternal sober color and black brings it down to zero. Incidentally, we can see the reason for this drastic and extreme reaction to black in the curve for dynamism and intimidation, where black generates the best dynamism and intimidation. We also observe an interesting relationship between "sober" and "glamorous" curves. Yellow and blue colors reduce the levels of sobriety with deep shades because with deep shades, the glamor quotient increases significantly. But it is the green and red, in that order, in their light shades which provide the best glamor. With dynamic curves, without

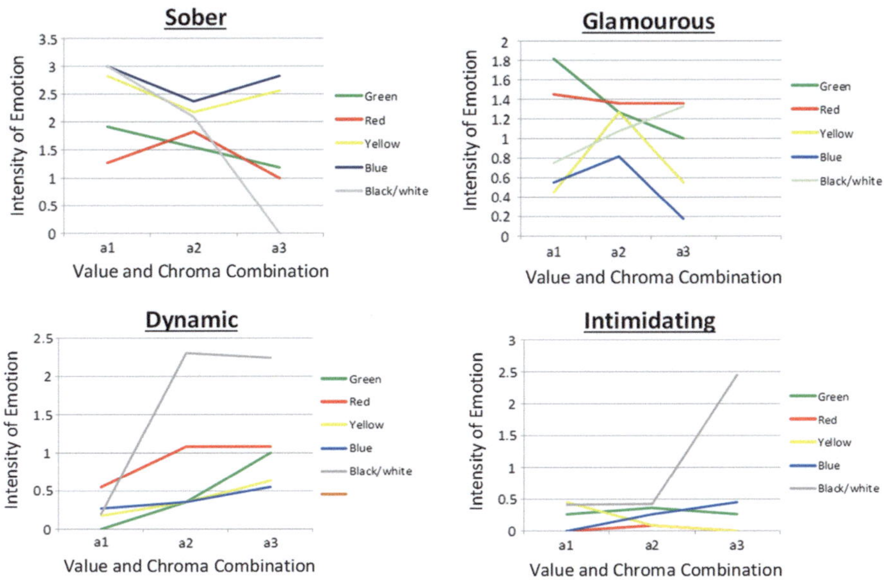

Fig. 10.4 Impact of mono-colors on emotions

exception the dark shade of each color provides the best dynamism, with black being the crown prince followed by dark red. With red, the deep shade is equally dynamic and gray provides the best quotient. Coming to the intimidating curves, black is by far the best and rest of the colors do not make much of a difference.

10.2 Combinations in a Single Primary Color

The second group of color plates is that of primary color shades in combination. These combinations are classified in two types—equal combinations (Fig. 10.5, left) and background–foreground combination (Fig. 10.5, right). The nomenclature for color and shade remains the same as explained for first group. Additionally, a dot between two colors a_1 and a_2 ($a_1.a_2$) signifies equal combination and a slash between the two (a_1/a_2) signifies a foreground of a_2 on a background a_1.

Similar to the previous group of colors, the SGDI score are plotted in Fig. 10.6 and wow, what a revelation! The games the color siblings play are mesmerizing. Read on.

The first thing we notice about these trends is that it broadly matches with the trend with single colors. This means shade combinations of blue dominate the sober curves, red and green, the glamor curves and red dominate both dynamic and intimidating curves. At a closer look, the light shades of blue as background with

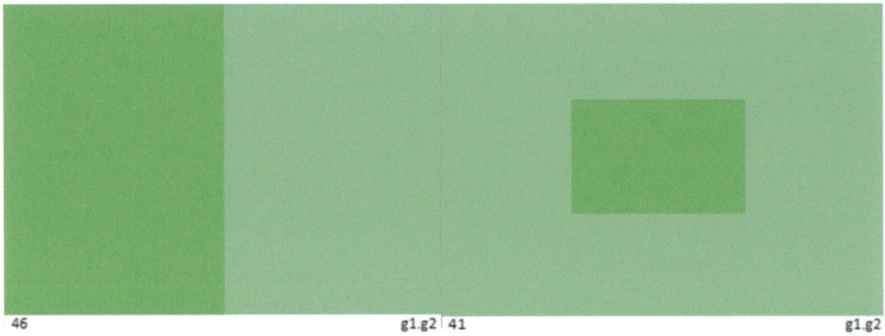

Fig. 10.5 Primary color combination for green

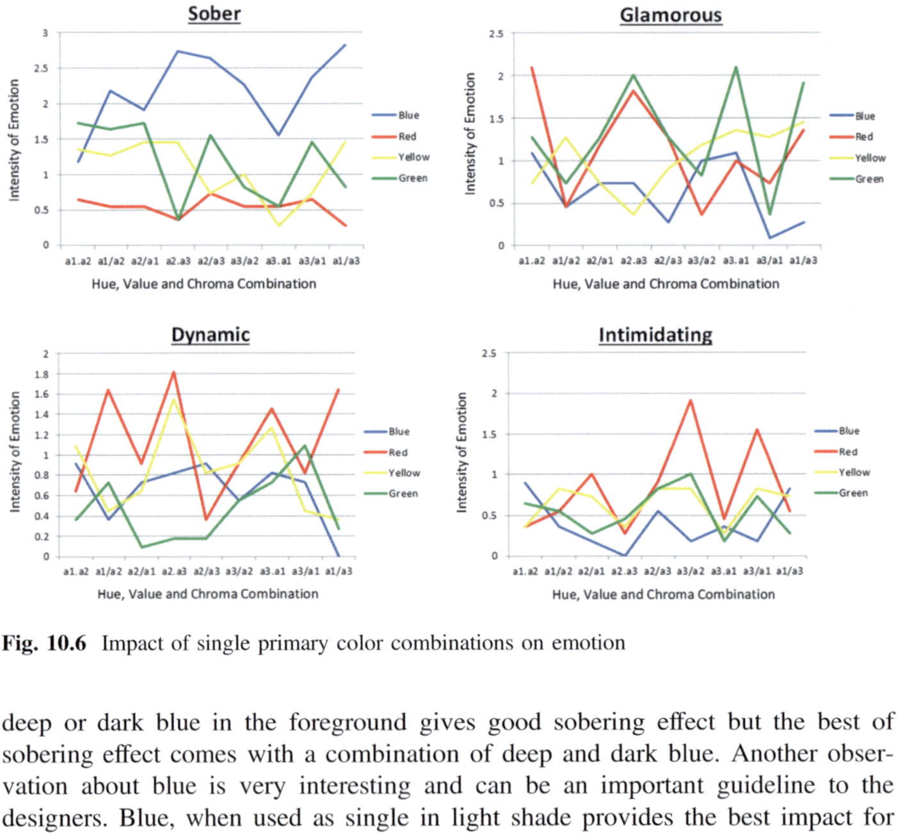

Fig. 10.6 Impact of single primary color combinations on emotion

deep or dark blue in the foreground gives good sobering effect but the best of sobering effect comes with a combination of deep and dark blue. Another observation about blue is very interesting and can be an important guideline to the designers. Blue, when used as single in light shade provides the best impact for sobriety at a score of 3. But if we need to use deep or dark shades, using as a combination of shades improves the impact. This can be observed if you compare Fig. 10.4 with Fig. 10.6. To a lesser degree, this is true for green as well. Any combination of red shades is to be strictly avoided for sober impact. So far as green is concerned, all the peaks are with background and foreground combinations with deep or dark shades in background.

If the emotion of glamor is to be aroused, red is the best choice with light and deep combination. Otherwise green with dark shade as one of the elements of combination generates the best glamor. With yellow, at best you can achieve some minor peaks and blue is to be avoided altogether if we want to achieve glamor by using combination of different shades with same color. Comparison of glamor curves in Figs. 10.4 and 10.6 leads us to similar conclusion as for sober curves that green and red colors enhance their impact when used as combination of shades.

Looking at the trends for dynamism, red dominates with dark and light combination or with deep shade in prominence. Comparison of dynamic curves in Figs. 10.4 and 10.6 leads us to similar conclusions as for sober and glamor trend. Red color when used as a combination of shades is more effective in generating dynamism than used a single shade, whether deep or dark. This observation is equally applicable for yellow and, to a lesser degree, blue.

We have a very surprising observation for intimidation curves. Red color is remarkably ineffective in creating intimidation when used a single shade. But, when used as a combination of shades, especially in background–foreground combination, red turns into the deadliest color.

10.3 Combinations of Different Primary Colors

Now, let us discuss the third group of plates—different colors, again in two types of combinations, i.e., equal and background–foreground (Fig. 10.7). We stick to the same nomenclature as in previous group which means letters signify color, numbers signify shade, dot between two colors a_1 and b_2 ($a_1.b_2$) signifies equal combination and a slash between the two (a_1/b_2) signifies foreground of b_2 on a background of a_1.

Look at the pictures (Figs. 10.8, 10.9, 10.10 and 10.11) and see the way colors team up to create a magic. But coming out of poetics and down to statistics, this

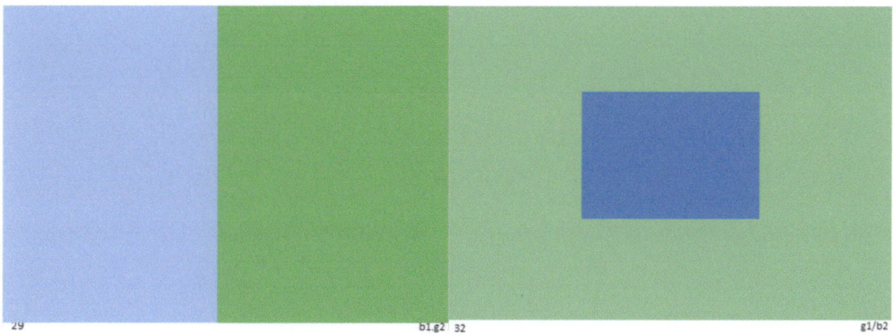

Fig. 10.7 Different primary color combinations

time we are in more complicated situation. First, the number of combination is more and second the combinations are more. As a result, each trend presents a challenge, which needs to be addressed separately. Let us see.

Figure 10.8 plots the curves for sober emotion and these curves bring out some remarkable revelations. While blue and yellow work best to generate the sober impact when light shades of both the colors are used, deep and dark combination produce the best effect when blue and green work as a team. Light shade figure on dark background works best with red-yellow and yellow-green combinations. Another astounding observation is that almost all peaks are with background–foreground combination, equal combination generates an insipid response in almost all cases.

A comparison with Figs. 10.4 and 10.6 suggests that blue as a color loses its sobering impact if used in combination with colors excepting yellow. Though in combination with green using deep and dark shades, it maintains its domination but the degree of impact is somewhat reduced. On the other hand if used in light shade in combination with light yellow, it is able to achieve the same degree of impact.

The picture for glamor (Fig. 10.9) is interesting indeed. All color combination can produce a glamor effect if used in specific shades and combinations. For example, red-yellow generates the best glamor in four combinations (light yellow picture on light red background, light red and dark yellow equal combination, light yellow picture on dark red background and dark red picture on dark yellow background), yellow-green in seven combinations, blue-red in three, blue-yellow in two, red-green in two, and blue-green in one combination.

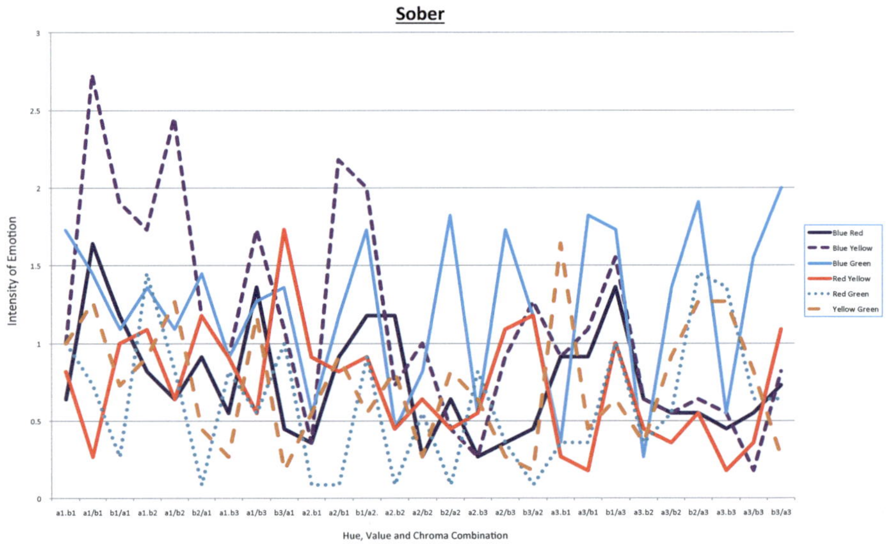

Fig. 10.8 Impact of different primary color combinations on sober emotion

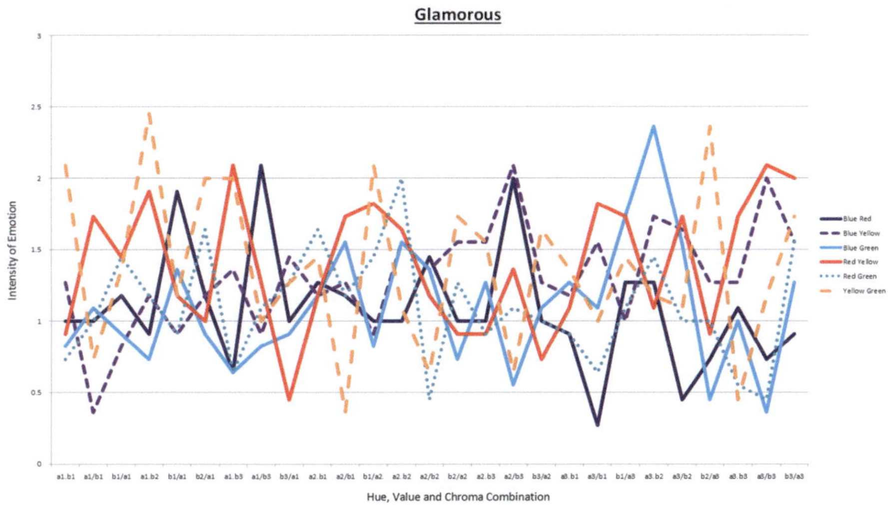

Fig. 10.9 Impact of different primary color combinations on glamor emotion

A study of peaks and valleys reveals a confusing picture as both show almost equal preference for equal and background–foreground combinations which is in contrast to sober trends where preference is clear. Another observation in contrast to sober trends is that colors used in combination broadly enhance the glamor impact in comparison to the case where they are used in isolation or shade combinations of the same color. For example, yellow and green when used in isolation are not able to exceed an intensity of 1.8 for glamor and even as combination of different shades of same color, the maximum they are able to achieve is an intensity of 2. But when used in combination with each other, they are able to reach the intensity of 2.5 in many cases. Similarly, blue as a single color creates an absolutely tepid response and even as combination of shades, it is lukewarm at best. But when used in combination with green it shows a huge improvement. So a broad statement can be safely made that colors used in combination enhance the glamor appeal of design.

Dynamism presents a relatively simpler picture (Fig. 10.10). Red and yellow combination largely dominates with light as one of the shades though blue-green and yellow-green also work best in some shade combinations. But story changes in deep-dark territory and blue-red combination rules here. Not only this, the intensity of dynamism also improves with deep-dark combinations. Peaks and valleys comparison does not reveal any specific preference for either equal combinations or background–foreground combination.

Comparison with single color and shade combination in single color does not surprise. Red maintains its dominance in all three cases (except gray and black, but they are not competing as of now) and the intensity of dynamism improves with red teaming up with other colors.

No clear picture emerges for intimidation (Fig. 10.11). Though yellow and green in equal combination with deep shades have the most intimidating score, it is the

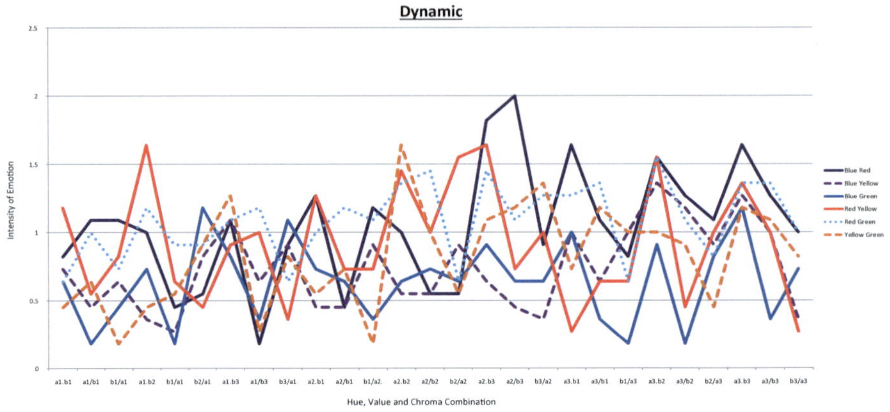

Fig. 10.10 Impact of different primary color combinations on dynamism emotion

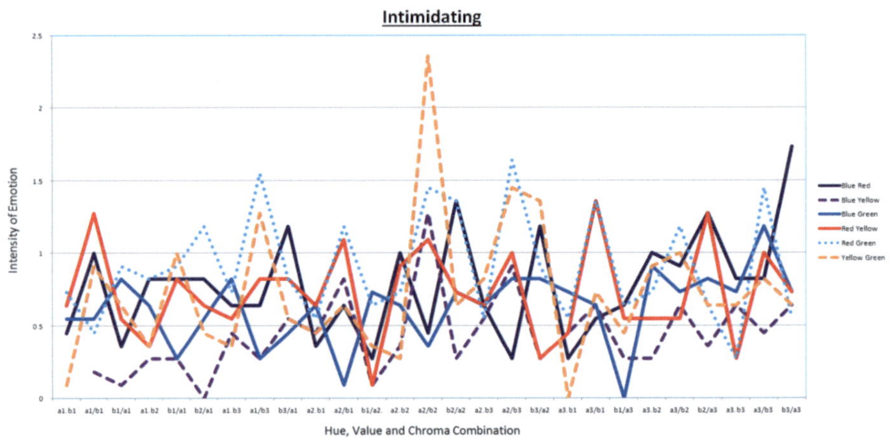

Fig. 10.11 Impact of different primary color combinations on intimidation emotion

red and green team which scores high with most shade combinations. Again, we can rule out any rule as to whether background–foreground combination should be preferred for intimidation or equal one.

Comparison of these curves with earlier intimidating curves yields two conclusions. Whereas, red generates the most desirable effect with single color shade combinations, it is yellow-green team which is most devastating as color combination, of course, in deep shades. Though, red maintains its presence in most other combinations. The second conclusion is rather insipid. The intensity, though more than single color (except black of course), remains rather tame in comparison to single color shade combinations (again, green-yellow deep shade equal combination is an exception).

10.4 Combination of Primary Colors, White, Black and Gray

The next set of color plates are for combination of colors with white (Fig. 10.12), black (Fig. 10.13) and gray (Fig. 10.14) and the combination of these three colors with each other. The first set consists of combination (both equal and background–foreground) of white with rest. The next set of diagrams shows the trends.

Here, the sober trends for combinations with white show that yellow and blue in their light shades create good effect. Gray and black are also equally good. Having said that, the fact remains that these colors are far more effective when used solo. Not only this, blue, when used as a combination of its shades is even more effective. Comparing this trend with the trend for other combinations reveals one more fact. The best teammate for blue is yellow and this pair has been unbeaten so far at least in doubles event. One more trend comes out clearly. By and large, equal combinations with white are far less effective in comparison to background–foreground combination. Sure, this is not news as we observed the same tendency earlier also.

For glamor also, the trend continues from color combination. We had earlier observed that glamor is created best when colors combine and among the combinations, red, and white works best. This combination of red and white is better than

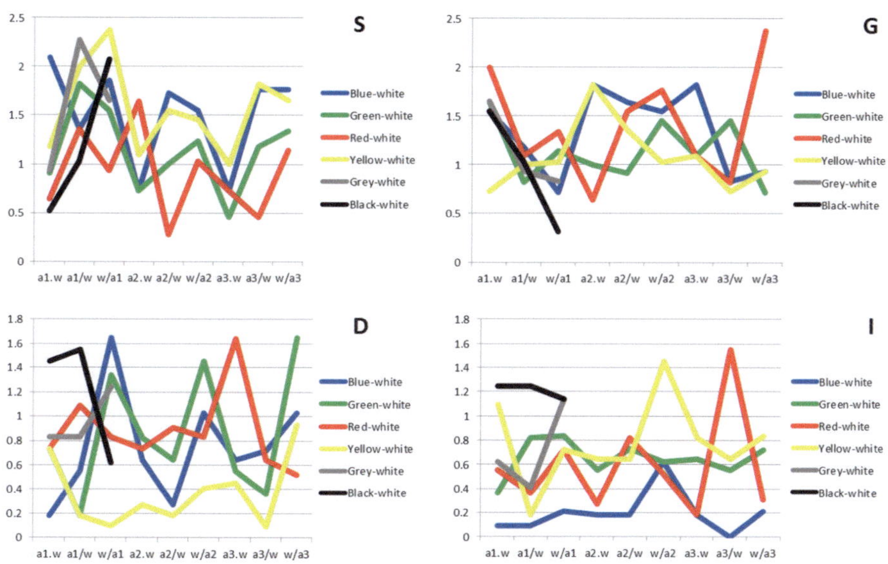

Fig. 10.12 Impact of combinations of white on emotion

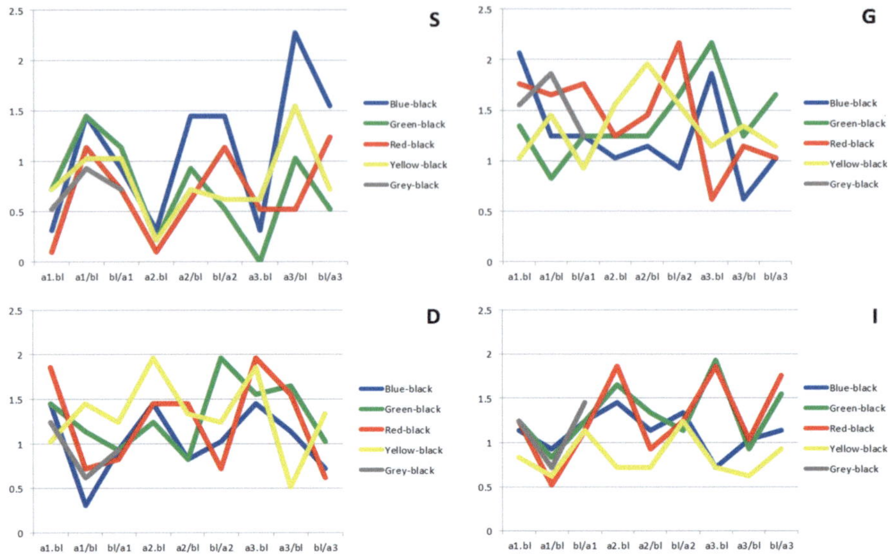

Fig. 10.13 Impact of combinations of black on emotion

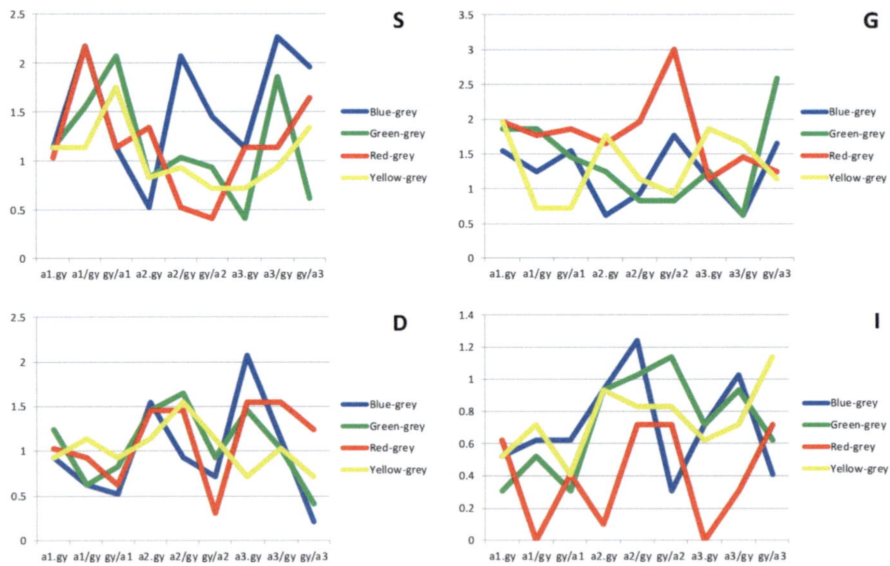

Fig. 10.14 Impact of combinations of gray on emotion

any other pair of red, but not better than green and yellow which perform equally well. Even among the red-white pairs, it is the dark red foreground on white background which works best. Another observation for glamor is that it has no special preference for any specific type of combination and all types of combination work equally well or equally bad.

For generating dynamism, if combining colors with white is not such an exciting idea, it is not such a bad idea as well. Except yellow, all colors generate an intensity up to 1.65 which roughly equal to the maximum value generated by either other combinations or color siblings or solo colors. Of course, black as solo is the best option so far for dynamism, whereas black in combination with white remains below par. And again, though you find no preference for either equal or background–foreground combinations, interestingly background–foreground combinations prefer a roller-coaster ride with both peaks and crests occupied by them and equal combinations preferring to be somewhere in the middle.

Creating intimidating effect by combining any color with white seems to be an avoidable idea. Clearly, you should choose other options like red siblings or simply using black or using yellow-green combination. Yes, if you want to eliminate intimidation altogether, using blue in combination with white will work quite well.

Combination of colors with black is rather lackluster in generating sober emotion. Whether you compare it with solo colors, color siblings or colors combining with each other, the intensity is rather low. It is low even in comparison to colors combining with white. The lone face-saver is blue, which in one of the combination is able to achieve a somewhat decent intensity. The "sober" emotion shows a clear preference for background–foreground combination, preferably with black as foreground.

The story turns slightly better with "glamor" curves. In broad terms, all colors improve their "glamor" intensity, when they combine with black. Though, the comparison with color siblings and colors combining with each other is not encouraging. Here, the trends do not show any preference for equal or background–foreground combination. The colors which create the best impact with black are red and green, red with black in the background and deep red in foreground and dark green in equal combination with black.

Combination with black works like a booster dose for "dynamic" curves. All colors show an improved intensity whether you compare with solo colors, color siblings or colors combing with each other. Moreover, they have a clear preference for black, if they have to choose between "black" and "white" as their partner to create dynamism. Another interesting facet with these curves is a broad preference for equal combinations. Inversely all crests are with background–foreground combination.

We have observed earlier that black has a propensity to intimidate and story is no different here. Like the dynamic curves, all colors improve their intimidation while combining with black, except for yellow which generates better intimidation with green and white. And like "dynamism", intimidation also works better with equal combination with crests preferring background–foreground in all cases.

So, the broader trends for combining with black show a general tendency of improvement for dynamism and intimidation, whereas "sober" and "glamor" do not respond enthusiastically. Moreover, a background–foreground combination works better with "sober" and "glamor" whereas equal combinations work better with "dynamism" and "intimidation".

With combinations with gray, so far as sober curves are concerned, gray is not badly placed. All colors except yellow are able to achieve intensities above 2 in combination with gray. Yellow, of course, acts best as solo or in combination with white whereas blue is best both as solo and as color siblings. Here again, peaks have a special preference for background–foreground combination. Though, you can create low sober quotient with both types of combinations.

For glamor, gray has special affinity for red with deep red in foreground. Dark green also works well with gray in the background achieving an intensity of 2.5 which is better than what any color achieves with black or white. Among other colors, only blue is never able to reach a level of 2, which is quite a feat. Peaks and valleys do not show any preference either for equal or for background–foreground combination.

Combinations with gray do not have much to talk about either for dynamism or for intimidation. Except for blue which is able to achieve a level of 2 for dynamism, maximum they are able to reach is 1.5. Yes, if a very low level of dynamism or intimidation is required, combinations with gray have many options.

Now considering all options together, we can choose the best options for generating an emotion. Table 10.2 provides a guideline for such a purpose.

Table 10.2 Survey result trends for color

Character	Rank	Sheet no	Score	Detail
Sober	1	b1	3.00	Light blue
	2	b1/b3	2.82	Dark blue on light blue background
	3	b3	2.82	Dark blue
	4	y3/r3	2.82	Dark red on a Dark yellow
Glamorous	1	gy/r2	3.00	Deep red on a gray background
	2	gy/g3	2.58	Dark green on a gray background
	3	b3.g2	2.36	Dark blue in equal combination with deep green
Dynamic	1	gy.b3	2.07	Dark blue and gray equal combination
	2	bl/g2	1.96	Deep green on a black background
	3	bl.r3	1.96	Dark red and black equal combination
Intimidating	1	y2/g1	2.64	Light green on deep yellow background
	2	y2/g2	2.36	Deep green on deep yellow background
	3	r3/r2	1.91	Deep red on Dark red background

10.5 Texture

To what extent does texture contribute to color in generating an emotion? Our study shows—a lot. In fact, it is almost equal. You can see yourself in the illustration (Fig. 10.15) how the impact changes with texture from plain to glossy to rough and it is not too difficult to imagine a car in all these three textures and feel the impact. But the bigger question is—do these three textures cover the whole gamut. The clear answer is no. Texture is after all the visual feel of the tactile feeling and there can be innumerable options. It is quantitative science and to cover the whole gamut is of course possible. But then, it will be research as big as this book, which in any case is not our objective. So I have been humble in my ambition to study the impact of texture and chose only three—the ones which are mostly applied on motorcycle surfaces (Fig. 10.15).

Another big issue is that it is not possible to separate the impact of texture from that of color. So, what I did was prepare plates of same texture with different colors and study the responses. Again, the colors chosen were the primary ones with light, deep, and dark shades in addition to black. For more realistic response I used real surfaces for chrome and silver as any photographic representation would have been unrealistic.

Let us come straight to results. For the sake of simplicity I have shown the mean value of emotional intensity for different shades of a color for showing the result here (otherwise the graph will be too big to get any real feel).

The result for sober emotion (Fig. 10.16) shows that by and large intensity is highest with glossy surface and chrome color. General trend is almost same for plain and rough except for white, where plain surface is more sober.

Results for glamor emotion (Fig. 10.17) indicate clearly that by and large, all types of surface generate almost same glamor except glossy chrome. And frankly, the glamor generated by other surfaces is not much to talk about.

Chrome color in glossy and silver in plain are more effective as far as dynamic emotion in concerned (see Fig. 10.18).

Figure 10.19 shows that chrome in glossy and black in rough and plain are more intimidating than others.

Fig. 10.15 The three textures

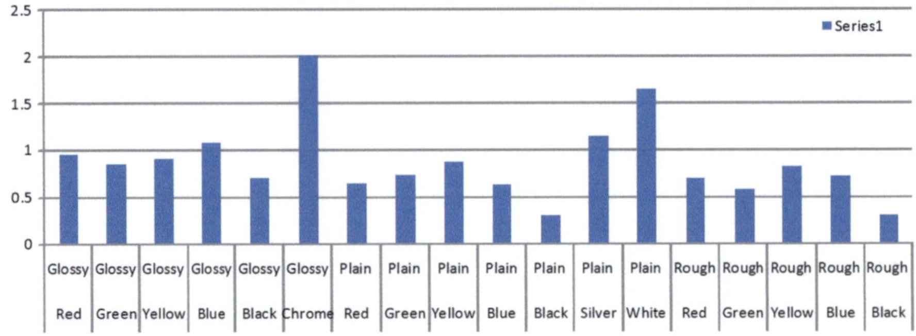

Fig. 10.16 Survey result trends for texture (S)

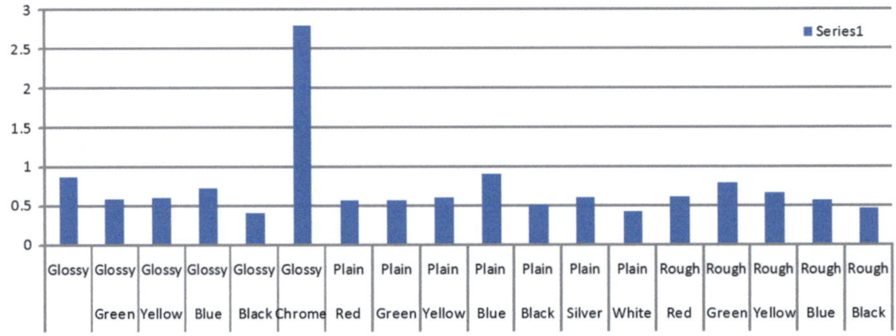

Fig. 10.17 Survey result trends for texture (G)

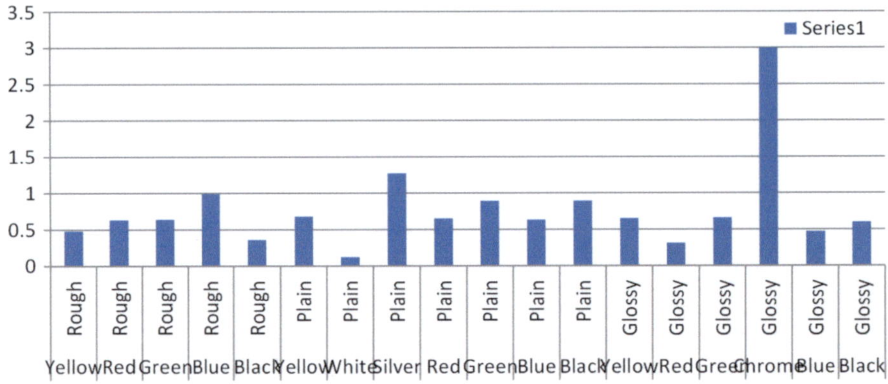

Fig. 10.18 Survey result trends for texture (D)

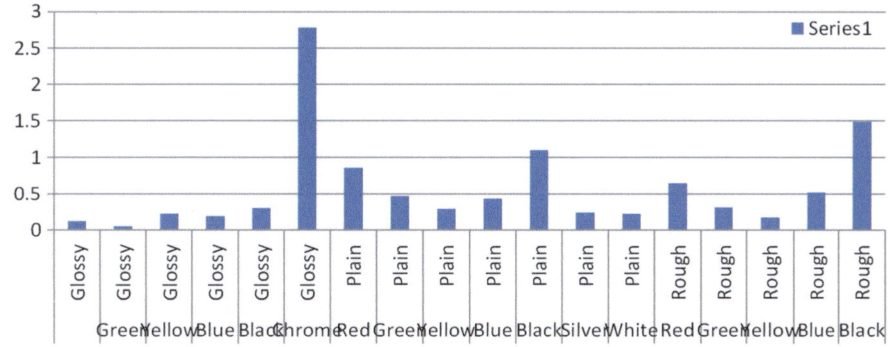

Fig. 10.19 Survey result trends for texture (I)

Table 10.3 Survey result trends for texture (mean)

	Glossy	Plain	Rough
S	1.04	0.78	0.68
G	0.7	0.72	0.66
D	0.82	0.61	0.64
I	0.34	0.51	0.49

The figures shown above and Table 10.3 clearly indicate that by and lage chrome plating generates the highest intensity for all kinds of emotions. If we exclude chrome plating, glossy surface generates best results for sober and dynamic characters, plain for glamor, and rough for intimidation.

We are all born with color as part of our consciousness since the moment of our birth and its relationship with emotions has been an integral part of our sensibility. Every expression of our sensibility expressed in form of literature and art is full of expressions of this relationship. But expressing this relationship in colorful (pun intended) words is one thing and having an objective and quantitative view of this relationship is quite another. But this quantitative reality is also a result of our consciousness built over centuries in the history of a particular society and therefore is not an absolute truth. For this very reason this relationship may change in a different societies and the quantitative relationship should be established again. But this disclaimer should not take us away from the fact this relationship can be expressed objectively and quantitatively in a specific context of a society.

Appendix 1: Emotion Scores for Color Combinations

	S	G	D	I
b3/b1	2.36	0.09	0.73	0.18
b3/b2	2.27	1.00	0.55	0.18
b2/b3	2.64	0.27	0.91	0.55
b2/b1	1.91	0.73	0.73	0.18
b1/b3	2.82	0.27	0.00	0.82
b1/b2	2.18	0.45	0.36	0.36
b2.b3	2.73	0.73	0.82	0.00
b1.b2	1.18	1.09	0.91	0.09
b1.b3	1.55	1.09	0.82	0.36
b3	2.82	0.18	0.55	0.45
b2	2.36	0.82	0.36	0.27
b1	3.00	0.55	0.27	0.00
b3/g3	1.55	0.36	0.36	1.18
b2/g3	1.73	0.55	0.64	0.82
b3/g2	1.36	1.64	0.18	0.73
b2/g2	0.82	1.36	0.73	0.36
b3/g1	1.82	1.09	0.36	0.64
b2/g1	1.18	1.55	0.64	0.09
b1/g1	1.45	1.09	0.18	0.55
b1/g2	1.09	1.36	0.18	0.27
b1/g3	1.27	0.82	0.36	0.27
b2.g1	0.55	1.18	0.73	0.64
b2.g2	0.45	1.55	0.64	0.64
b2.g3	0.55	1.27	0.91	0.64
b3.g1	0.36	1.27	1.00	0.73
b3.g2	0.27	2.36	0.91	0.91
b3.g3	0.55	1.00	1.18	0.73
b1.g1	1.73	0.82	0.64	0.55
b1.g2	1.36	0.73	0.73	0.64
b1.g3	0.91	0.64	0.82	0.82
g1/b1	1.09	0.91	0.45	0.82
g1/b2	1.73	0.82	0.36	0.73
g1/b3	1.73	1.73	0.18	0.00
g2/b1	1.45	0.91	1.18	0.55
g2/b2	1.82	0.73	0.64	0.73
g2/b3	1.91	0.45	0.82	0.82
g3/b1	1.36	0.91	1.09	0.45
g3/b2	1.18	1.09	0.64	0.82
g3/b3	2.00	1.27	0.73	0.73
g1/g3	0.82	1.91	0.27	0.27
g1/g2	1.64	0.73	0.73	0.55
g2/g1	1.73	1.27	0.09	0.27
g2/g3	1.55	1.27	0.18	0.82
g3/g1	1.45	0.36	1.09	0.73
g3/g2	0.82	0.82	0.55	1.00
g1.g2	1.73	1.27	0.36	0.64
g2.g3	0.36	2.00	0.18	0.45
g3.g1	0.55	2.09	0.73	0.18
g1	1.91	1.82	0.00	0.27
g2	1.55	1.27	0.36	0.36
g3	1.18	1.00	1.00	0.27
b1/w	1.36	1.18	0.55	0.09
b2/w	1.73	1.64	0.27	0.18
b1.w	2.09	1.55	0.18	0.09
b2.w	0.73	1.82	0.64	0.18
g3/w	1.18	1.45	0.36	0.55
g2/w	1.00	0.91	0.64	0.73
g1/w	1.82	0.82	0.18	0.82
g3.w	0.45	1.09	0.55	0.64
g2.w	0.73	1.00	0.82	0.55
g1.w	0.91	1.64	0.73	0.36
g3/r1	1.00	1.27	0.64	0.82
g2/r1	0.09	1.64	0.91	1.18
g1/r1	0.27	1.45	0.73	0.91
g3/r2	0.09	1.00	1.27	0.91
g2/r2	0.09	1.27	0.64	1.36
g1/r2	0.91	1.45	1.09	0.64
g3/r3	0.64	1.55	1.00	0.55
g2/r3	1.45	1.00	0.82	0.64

	S	G	D	I
g3.r2	0.82	0.91	1.45	0.55
g3.r3	1.36	0.55	1.36	0.27
g2.r1	1.45	1.18	1.18	0.82
g2.r2	0.09	2.00	1.36	0.73
g2.r3	0.36	1.45	1.55	0.73
g1.r1	1.00	0.73	0.64	0.73
g1.r2	0.09	1.64	1.00	0.55
g1.r3	0.36	0.91	1.27	0.55
r3/g1	0.36	0.64	1.36	1.36
r3/g2	0.55	1.00	1.09	1.18
r3/g3	0.64	0.45	1.36	1.45
r2/g1	0.09	1.18	1.18	1.18
r2/g2	0.55	0.45	1.45	1.45
r2/g3	0.36	1.09	1.09	1.64
r1/g1	0.73	1.00	1.00	0.45
r1/g2	0.82	0.91	0.91	0.91
r1/g3	0.55	1.09	1.18	0.55
b1/r1	1.64	1.00	1.09	1.00
b1/r2	0.64	1.91	0.45	0.82
b1/r3	1.36	2.09	0.18	0.64
b2/r1	0.91	1.18	0.45	0.64
b2/r2	0.27	1.45	0.55	0.45
b2/r3	0.36	2.00	1.00	0.27
b3/r1	0.91	0.27	1.09	0.55
b3/r2	0.55	0.45	1.27	0.91
b3/r3	0.55	0.73	1.27	0.82
b1.r1	0.64	1.00	0.82	0.45
b1.r2	0.82	0.91	1.00	0.82
b1.r3	0.55	0.64	1.09	0.64
b2.r1	0.36	1.27	1.27	0.36
b2.r2	1.18	1.00	1.00	1.00
b2.r3	0.27	1.00	1.82	0.64
b3.r1	0.91	0.91	1.64	0.27
b3.r2	0.64	1.27	1.55	1.00
b3.r3	0.45	1.09	1.64	0.82
r1/b1	1.18	1.18	1.09	0.36
r1/b2	1.18	1.00	1.18	0.27
r1/b3	0.00	0.11	0.03	0.07
r2/b1	0.00	0.18	0.03	0.07
r2/b2	0.00	0.07	0.01	0.02
r2/b3	0.00	0.15	0.03	0.05
r3/b1	0.00	0.16	0.00	0.02
r3/b2	0.00	0.14	0.01	0.07
r3/b3	0.00	0.13	0.05	0.07
r3/r1	0.00	0.07	0.02	0.09
r3/r2	0.00	0.16	0.02	0.06
r3/w	0.00	0.03	0.01	0.04
r2/r1	0.00	0.05	0.03	0.02
r2/r3	0.00	0.17	0.00	0.02
r2/w	0.00	0.14	0.02	0.03
r1/r3	0.00	0.11	0.05	0.02
r1/r2	0.00	0.12	0.02	0.01
r1/w	0.00	0.16	0.01	0.02
r1.r2	0.00	0.19	0.01	0.01
r2.r3	0.00	0.07	0.03	0.02
r3.r1	0.00	0.11	0.02	0.05
r1.w	0.00	0.09	0.03	0.07
r2.w	0.00	0.17	0.01	0.07
r3.w	0.00	0.04	0.02	0.06
r1	0.00	0.07	0.04	0.05
r2	0.00	0.08	0.03	0.03
r3	0.00	0.09	0.03	0.07
b1/y1	0.00	0.01	0.04	0.11
b1/y2	0.00	0.02	0.03	0.08
b1/y3	0.00	0.01	0.06	0.08
b2/y1	0.00	0.01	0.03	0.12
b2/y2	0.00	0.08	0.05	0.06
b2/y3	0.00	0.06	0.05	0.05
b3/y1	0.00	0.13	0.04	0.06
b3/y2	0.00	0.00	0.00	0.00
b3/y3	0.00	0.00	0.00	0.00
b1.y1	0.00	0.00	0.00	0.00
b1.y2	0.00	0.00	0.00	0.00
b1.y3	0.00	0.00	0.00	0.00
b2.y1	0.00	0.00	0.00	0.00
b2.y2	0.00	0.00	0.00	0.00

	S	G	D	I
r2/b2	0.64	1.00	0.55	1.36
r2/b3	0.55	0.73	1.09	1.27
r3/b1	0.45	1.00	0.91	1.18
r3/b2	0.73	0.73	1.09	1.27
r3/b3	0.73	0.91	1.00	1.73
r3/r1	0.64	0.73	0.82	1.55
r3/r2	0.55	0.36	0.91	1.91
r3/w	0.45	0.82	0.64	1.55
r2/r1	0.55	1.18	0.91	1.00
r2/r3	0.73	1.27	0.36	0.91
r2/w	0.27	1.55	0.91	0.82
r1/r3	0.27	1.36	1.64	0.55
r1/r2	0.55	0.45	1.64	0.55
r1/w	1.36	1.09	1.09	0.36
r1.r2	0.64	2.09	0.64	0.36
r2.r3	0.36	1.82	1.82	0.27
r3.r1	0.55	1.00	1.45	0.45
r1.w	0.64	2.00	0.73	0.55
r2.w	1.64	0.64	0.73	0.27
r3.w	0.73	1.09	1.64	0.18
r1	1.27	1.45	0.55	0.00
r2	1.82	1.36	1.09	0.09
r3	1.00	1.36	1.09	0.00
b1/y1	2.73	0.36	0.45	0.09
b1/y2	2.45	0.91	0.27	0.00
b1/y3	1.73	0.91	0.64	0.55
b2/y1	2.18	1.27	0.45	0.09
b2/y2	1.00	1.36	0.55	0.27
b2/y3	0.91	2.09	0.45	0.27
b3/y1	1.09	1.55	0.64	0.27
b3/y2	0.55	1.64	1.18	0.36
b3/y3	0.18	2.00	1.00	0.45
b1.y1	1.00	1.27	0.73	0.18
b1.y2	1.73	1.18	0.36	0.27
b1.y3	0.91	1.36	1.09	0.27
b2.y1	0.36	1.18	0.45	0.82
b2.y2	0.73	1.55	0.55	1.27
b2.y3	0.27	1.55	0.64	0.91
b3.y1	0.91	1.18	1.00	0.64
b3.y2	0.64	1.73	1.36	0.64
b3.y3	0.55	1.27	1.27	0.45
y1/b1	1.91	0.82	0.64	0.27
y1/b2	2.00	0.91	0.91	0.36
y1/b3	1.55	1.00	1.00	0.27
y2/b1	1.18	1.18	0.82	0.45
y2/b2	0.45	1.55	0.91	0.55
y2/b3	0.64	1.27	0.91	0.64
y3/b1	1.09	1.45	0.91	0.45
y3/b2	1.27	1.27	0.36	0.45
y3/b3	0.82	1.55	0.36	0.64
y1/w	2.00	1.00	0.18	0.18
y2/w	1.55	1.36	0.18	0.64
y3/w	1.82	0.73	0.09	0.64
b1.w	1.18	0.73	0.73	1.09
y2.w	1.09	1.82	0.27	0.64
y3.w	1.00	1.09	0.45	0.82
y1/y2	1.27	1.27	0.45	0.82
y1/y3	1.45	1.45	0.36	0.73
y2/y1	1.45	0.73	0.64	0.73
y2/y3	0.73	0.91	0.82	0.82
y3/y1	0.73	1.27	0.45	0.82
y3/y2	1.00	1.18	0.91	0.82
y1.y2	1.36	0.73	1.09	0.36
y2.y3	1.45	0.36	1.55	0.36
y3.y1	0.27	1.36	1.27	0.27
g1/y1	0.73	1.36	0.18	0.64
g1/y2	0.55	2.09	0.18	0.36
g1/y3	0.64	1.45	1.00	0.45
g2/y1	0.45	2.00	0.91	0.45
g2/y2	0.82	1.73	0.55	0.64
g2/y3	1.27	2.36	0.45	0.64
g3/y1	0.18	1.27	0.82	0.55
g3/y2	0.18	1.64	1.36	0.36
g3/y3	0.27	1.73	0.82	0.64
g1.y2	0.55	1.45	0.55	0.45
g1.y1	1.00	2.09	0.45	0.09

	S	G	D	I
g2.y2	0.00	0.06	0.02	0.12
g2.y3	0.00	0.05	0.05	0.12
g3.y1	0.00	0.04	0.04	0.11
g3.y2	0.00	0.07	0.05	0.12
g3.y3	0.00	0.07	0.05	0.16
y1/g1	0.00	0.06	0.04	0.14
y1/g2	0.00	0.05	0.04	0.17
y1/g3	0.00	0.04	0.03	0.14
y2/g1	0.00	0.05	0.04	0.09
y2/g2	0.00	0.07	0.02	0.08
y2/g3	0.00	0.02	0.04	0.07
y3/g1	0.00	0.02	0.07	0.05
y3/g2	0.00	0.05	0.07	0.05
y3/g3	0.00	0.12	0.05	0.03
r1/y1	0.00	0.06	0.03	0.03
r1/y2	0.00	0.03	0.08	0.02
r1/y3	0.00	0.05	0.07	0.04
r2/y1	0.00	0.06	0.03	0.05
r2/y2	0.00	0.15	0.03	0.02
r2/y3	0.00	0.07	0.07	0.02
r3/y1	0.00	0.12	0.02	0.00
r3/y2	0.00	0.17	0.05	0.01
r3/y3	0.00	0.09	0.05	0.00
r1.y1	0.00	0.25	0.02	0.01
r1.y2	0.00	0.22	0.01	0.00
r1.y3	0.00	0.16	0.03	0.05
r2.y1	0.00	0.20	0.02	0.01
r2.y2	0.00	0.09	0.02	0.02
r2.y3	0.00	0.08	0.02	0.02
r3.y1	0.00	0.10	0.03	0.02
r3.y2	0.00	0.05	0.05	0.03
r3.y3	0.00	0.02	0.05	0.04
y1/r1	0.00	0.09	0.03	0.02
y1/r2	0.00	0.16	0.02	0.02
y1/r3	0.00	0.08	0.05	0.02
y2/r1	0.00	0.03	0.02	0.07
y2/r2	0.00	0.07	0.02	0.12
y2/r3	0.00	0.02	0.03	0.08
y3/r1	0.00	0.08	0.05	0.06
y3/r2	0.00	0.06	0.06	0.06
y3/r3	0.00	0.05	0.06	0.04
y1	0.00	0.17	0.03	0.02
y2	0.00	0.18	0.04	0.03
y3	0.00	0.14	0.05	0.02

	S	G	D	I
b3/w	1.76	0.83	0.72	0.00
b3/bl	1.65	0.31	0.93	1.14
b3/gy	2.27	0.62	1.14	1.03
b2/bl	1.45	1.14	0.83	1.14
b2/gy	2.07	0.93	0.93	1.24
b1/bl	1.45	1.24	0.31	0.93
b1/gy	2.17	1.24	0.62	0.62
g3/bl	1.03	1.24	1.65	0.93
g3/gy	1.86	0.62	1.03	0.93
g2/bl	0.93	1.24	0.83	1.34
g2/gy	1.03	0.83	1.65	1.03
g1/bl	1.45	0.83	1.14	0.83
g1/gy	1.55	1.86	0.62	0.52
bl/w	1.03	1.03	1.55	1.24
bl/gy	0.72	1.24	0.93	1.45
r3/bl	0.52	1.14	1.55	1.03
r3/gy	1.14	1.45	1.55	0.31
r2/bl	0.62	1.45	1.45	0.93
r2/gy	0.52	1.96	1.45	0.72
r1/bl	1.14	1.65	0.72	0.52
r1/gy	2.17	1.76	0.93	0.00
y3/bl	1.55	1.34	0.52	0.62
y3/gy	0.93	1.65	1.03	0.72
y2/bl	0.72	1.96	1.34	0.72
y2/gy	0.93	1.14	1.55	0.83
y1/bl	1.03	1.45	1.45	0.62
gy/bl	0.93	1.86	0.62	0.72
w/bl	2.07	0.31	0.62	1.14
gy/w	2.27	0.93	0.83	0.41
w/bl	1.65	0.83	1.24	1.14
y1/gy	1.14	0.72	1.14	0.72
w/b3	1.76	0.93	1.03	0.21
w/b2	1.55	1.55	1.03	0.62
w/b1	1.86	0.72	1.65	0.21
w/g3	1.34	0.72	1.65	0.72
w/g2	1.24	1.45	1.45	0.62
w/g1	1.55	1.14	1.34	0.83
w/r3	1.14	2.38	0.52	0.31
w/r2	1.03	1.76	0.83	0.52
w/r1	0.93	1.34	0.83	0.72
w/y3	1.65	0.93	0.93	0.83
w/y2	1.45	1.03	0.41	1.45

	S	G	D	I
bl/b2	1.45	0.93	1.03	1.34
bl/b1	0.93	1.24	0.93	1.24
bl/g3	0.52	1.65	1.03	1.55
bl/g2	0.52	1.65	1.96	1.14
bl/g1	1.14	1.24	0.93	1.24
bl/r3	1.24	1.03	0.62	1.76
bl/r2	1.14	2.17	0.72	1.24
bl/r1	0.72	1.76	0.83	1.14
bl/y3	0.72	1.14	1.34	0.93
bl/y2	0.62	1.55	1.24	1.24
bl/y1	1.03	0.93	1.24	1.14
gy/b3	1.96	1.65	0.21	0.41
gy/b2	1.45	1.76	0.72	0.31
gy/b1	1.14	1.55	0.52	0.62
gy/g3	0.62	2.58	0.41	0.62
gy/g2	0.93	0.83	0.93	1.14
gy/g1	2.07	1.45	0.83	0.00
gy/r3	1.65	1.24	1.24	0.72
gy/r2	0.41	3.00	0.31	0.72
gy/r1	1.14	1.86	0.62	0.41
gy/y3	1.34	1.14	0.72	1.14
gy/y2	0.72	0.93	1.14	0.83
gy/g1	1.76	0.72	0.93	0.41
bl.b3	0.31	1.86	1.45	0.72
bl.b2	0.31	1.03	1.45	1.45
bl.b1	0.31	2.07	1.45	1.14
bl.g3	0.00	2.17	1.55	0.93
bl.g2	0.21	1.24	1.24	1.65
bl.g1	0.72	1.34	1.45	1.24
bl.r3	0.52	0.62	1.96	1.86
bl.r2	0.10	1.24	1.45	1.86
bl.r1	0.10	1.76	1.86	1.24
bl.y3	0.62	1.14	1.86	0.72
bl.y1	0.72	1.03	1.03	0.83
bl.y2	0.21	1.55	1.96	0.72
bl.w	0.52	1.55	1.45	1.24
bl.gy	0.52	1.55	1.24	1.24
gy.b3	1.14	1.14	2.07	0.72
gy.b2	0.52	0.62	1.55	0.93
gy.b1	1.14	1.55	0.93	0.52
gy.g3	0.41	1.45	1.24	0.72
gy.g2	0.83	1.24	1.45	0.93
gy.g1	1.14	1.86	1.24	0.31
gy.r3	1.14	1.14	1.55	0.00
gy.r2	1.34	1.65	1.45	0.10
gy.g1	1.03	1.96	1.03	0.62
gy.y3	0.72	1.86	0.72	0.62
gy.y2	0.83	1.76	1.14	0.93
gy.y1	1.14	1.96	0.93	0.52
gy.w	0.93	1.65	0.83	0.62

		S	G	D	I
w	1	2.99	0.75	0.20	0.41
bl	2	0.00	1.33	2.24	2.45
gy	3	2.09	1.08	2.31	0.43

Appendix 2: Emotion Scores for Color and Texture Combinations

		Hue	Value	Chroma	Texture		X=x'/2.29*3
1		Red	Medium	Low	Plain	S	0.71
						G	0.77
						D	0.51
						I	0.66
2		Red	Medium	Low	Glossy	S	1.10
						G	0.92
						D	0.09
						I	0.09
3		Red	Medium	Low	Rough	S	0.83
						G	0.51
						D	0.33
						I	0.80
4	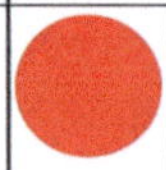	Red	Medium	High	Plain	S	0.45
						G	0.48
						D	1.10
						I	0.89
5		Red	Medium	High	Glossy	S	0.60
						G	1.49
						D	0.48
						I	0.00
6		Red	Medium	High	Rough	S	0.36
						G	0.66
						D	1.40
						I	0.45
7		Red	Low	Low	Plain	S	0.77
						G	0.45
						D	0.33
						I	1.04
8		Red	Low	Low	Glossy	S	1.16
						G	0.21
						D	0.36
						I	0.27
9		Red	Low	Low	Rough	S	0.89
						G	0.66
						D	0.15
						I	0.68
10		Green	Medium	Low	Plain	S	0.54
						G	0.83
						D	0.68
						I	0.30
11		Green	Medium	Low	Glossy	S	0.86
						G	0.45
						D	0.71
						I	0.06
12		Green	Medium	Low	Rough	S	0.83
						G	0.60
						D	0.57
						I	0.18
13		Green	Medium	High	Plain	S	0.30
						G	0.45
						D	1.31
						I	0.86
14		Green	Medium	High	Glossy	S	1.04
						G	0.51
						D	0.71
						I	0.00

		Hue	Value	Chroma	Texture		X=x'/2.29*3
15		Green	Medium	High	Rough	S	0.33
						G	1.07
						D	0.86
						I	0.30
16		Green	Low	Low	Plain	S	1.37
						G	0.42
						D	0.68
						I	0.24
17	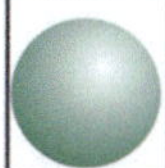	Green	Low	Low	Glossy	S	0.66
						G	0.77
						D	0.54
						I	0.09
18		Green	Low	Low	Rough	S	0.57
						G	0.71
						D	0.48
						I	0.45
19	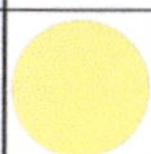	Yellow	Medium	Low	Plain	S	0.86
						G	0.36
						D	0.77
						I	0.18
20		Yellow	Medium	Low	Glossy	S	1.13
						G	0.21
						D	0.86
						I	0.06
21		Yellow	Medium	Low	Rough	S	1.01
						G	0.66
						D	0.33
						I	0.00
22	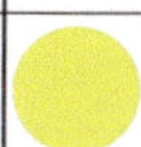	Yellow	Medium	High	Plain	S	1.01
						G	0.77
						D	0.86
						I	0.24
23		Yellow	Medium	High	Glossy	S	0.60
						G	1.13
						D	0.86
						I	0.12
24		Yellow	Medium	High	Rough	S	0.42
						G	1.13
						D	0.54
						I	0.21
25		Yellow	Low	Low	Plain	S	0.77
						G	0.66
						D	0.42
						I	0.45
26		Yellow	Low	Low	Glossy	S	1.04
						G	0.42
						D	0.21
						I	0.48
27		Yellow	Low	Low	Rough	S	1.07
						G	0.21
						D	0.57
						I	0.30
28		Blue	Medium	Low	Plain	S	0.27
						G	1.37
						D	0.66
						I	0.33

		Hue	Value	Chroma	Texture		X=x'/2.29*3
29		Blue	Medium	Low	Glossy	S	1.07
						G	0.68
						D	0.42
						I	0.18
30		Blue	Medium	Low	Rough	S	0.57
						G	0.77
						D	1.10
						I	0.30
31	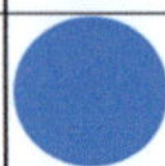	Blue	Medium	High	Plain	S	0.92
						G	0.92
						D	0.86
						I	0.21
32	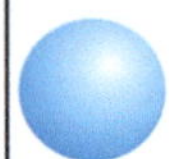	Blue	Medium	High	Glossy	S	0.92
						G	0.98
						D	0.71
						I	0.12
33		Blue	Medium	High	Rough	S	1.01
						G	0.42
						D	1.04
						I	0.33
34		Blue	Low	Low	Plain	S	0.68
						G	0.42
						D	0.36
						I	0.74
35	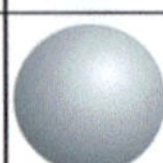	Blue	Low	Low	Glossy	S	1.25
						G	0.51
						D	0.27
						I	0.27
36		Blue	Low	Low	Rough	S	0.60
						G	0.48
						D	0.83
						I	0.92
37		Black			Plain	S	0.30
						G	0.51
						D	0.89
						I	1.10
38		Black			Glossy	S	0.71
						G	0.39
						D	0.60
						I	0.30
39		Black			Rough	S	0.30
						G	0.45
						D	0.36
						I	1.49
40		Chrome				S	2.02
						G	2.80
						D	3.00
						I	2.78
41		Silver				S	1.15
						G	0.59
						D	1.27
						I	0.24
42		White				S	1.65
						G	0.41
						D	0.12
						I	0.22

Chapter 11
Unity and Dynamism: The Interplay

Can someone describe, objectively that is, what makes the Taj Mahal the excellent work of architecture that it is? The answer is an emphatic YES.

The parameters of design for a desired narrative have been the crux of heated debates between practitioners and theorists of design. At the outset, two major theories have been dominating this debate—Gestalt theory and structuralism. Gestalt theory is based on Gestalt psychology, which believes that in visual human perception, the whole is bigger than the sum of its parts and hence the overall perception of design cannot be deconstructed into various elements. In Pratt's [1] opinion, it is actually the subversion of the way the eye and mind actively manipulate what we see, that causes something to be aesthetically pleasing. There are various Gestalts like Gestalt of closure, proximity, etc., which provide the additional meaning, which cannot be realized if deconstructed. Way back in 1941, Campbell [2] argued against the contention that the effect of the whole over the sum of the parts is negligible. On the other hand, structuralist thinkers like Kotnik [3] define structure as a pattern resulting from the network of formal relationships between the elements of a set and a topological structure like design is about the geometric organization of elements. To put it simply, the difference between the whole and the sum of the parts is also a mathematical construct and can be expressed in terms of mathematical relationships. Durfeld [4] defines structuralism as a rule-based design process translated into mathematical algorithms and resolved. This takes us to Derrida's deconstruction theory, which has been extensively applied in architecture but hardly in machine design. The basic hypothesis of this theory is that there is nothing beyond the text, and the sub-text (or the gestalten) is also a construct of the text, which can be deconstructed through the relationship and interdependence between (a) the binary of words, i.e., rich–poor, strong–weak, etc., and (b) the signifier and the signified (the symbol and its meaning). But the dilemma of the structuralist approach lies in its inability to understand the process of transformation [5]. There have been attempts to resolve this dilemma with the help of ornamentality through rhythm. These attempts, though all these debates took place in the realm of architecture and urban planning, can be particularly seen

S. Chandra, *Aesthetics: Quantification and Deconstruction*,
Design Science and Innovation, https://doi.org/10.1007/978-981-10-6235-3_11

in the design architecture of motorcycles, where a conscious attempt is made to create rhythm through the monotony of a large number of components. The reality or the ultimate solution, as always lies in the synthesis of these two extreme positions. As we discussed earlier, Gestalts themselves can be treated as mathematical constructs and algorithms generated to understand the process of transformation.

Now the question is, what are the gestalts which define the difference between the whole and the sum of many deconstructed elements like form, color, context, etc.. Let us look for the answer in the two pictures in Fig. 11.1. The first image is that of a Ferrari and we know it is a sports car. But let us pretend that we have not heard of Ferrari and what it means. Still, the image tells that it is a fast, real fast car. The other image is also that of a car. Very pleasing to the senses, but does not promise speed. Now, what is it that speaks for the images? No, it is neither their color nor their shape. It is something else. Let us understand this mystery through two terms that transcend beyond the individual parts—unity and dynamism.

11.1 Unity

Let us look at another two pictures in Fig. 11.2. The celebrated work by Vincent Van Gogh speaks volumes about how this whole universe including the stars, the galaxies, the earth, the trees, the mountains, and the human habitation is one single piece of design. In fact, all the elements of unity of design in automobiles are explained beautifully in this work. Similarly, the celebrated monument of love has this overarching theme of love dissolving into spirituality. This wholesome feel comes from the unity of form and lines combined with an ethereal continuity.

Philosophers of design like Young [6] and Elam [7] have described the unity of design as a significant visual factor. McClurg-Genevese [8] defines unity as a concept that describes the relationship between individual parts and the whole of a composition. Jirousek [9] defines it as an underlying principle that refers to the coherence of the whole, the sense that all of the parts are working together to

Fig. 11.1 Two cars. What makes the images different?

Fig. 11.2 "Starry Night" by Vincent Van Gogh and Taj Mahal. *Note* The unity of form and continuity

achieve a common result. Lauer et al. [10] define unity of design as a concept wherein a congruity or agreement exists among the elements in a design; as though a visual connection beyond mere chance has caused them to come together. Zelanski et al. [11] define it as a property of art or design, where the whole will seem to be more than the sum of its parts. But, for a moment, keep the definitions aside and think of a poem, where all lines are beautifully written but do not relate at all to each other. Or, let us make the problem a bit more complicated. Let us combine the most beautiful lines of Shakespeare, Wordsworth, Brecht, and Neruda to create one single poem. Will it make sense? You know the answer (what non-sense!) and the reason as well. Let us list down the reasons:

a. The lines should generate a sense of oneness. This means that all lines should convey the same emotions in the same context.
b. All lines should relate to each other. This means that the second line should start where the first line ends.
c. All lines should be part of the same flow.
d. All lines taken together should create one overarching theme.

Now, we transport this reasoning to the world of motorcycle design and explore the characteristics of the unity of motorcycle design. Thus, unity in the context of the motorcycle can be defined as the mutual compatibility between five factors—unity of form, continuity of parts, character lines, feature lines, and graphics:

a. Unity of form: Two objects are having unity of form if they have either similar or complimentary forms. Forms, as we know are either geometric or organic and can have surfaces with positive–positive or positive–negative spaces. Besides the similarity or complementation should be visibly distinct. The illustration in Fig. 11.3a is a good example of the unity of form where all elements are having different sizes and colors but are having the same form. Look at the motorcycle in Fig. 11.3b. The fuel tank, the side cover, and the rear cover, all have convex curvatures joined by sharp straight lines. This generates the unity of form in the motorcycle.

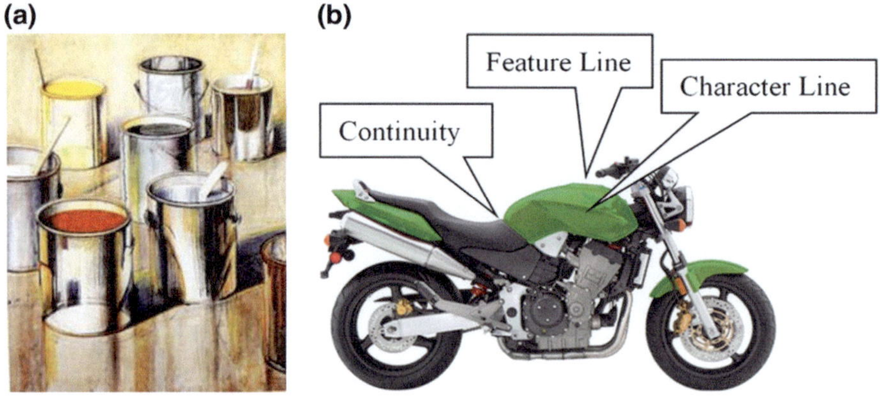

Fig. 11.3 **a** Unity of form, **b** continuity, character lines and feature lines in a motorcycle

b. Continuity: When two objects appear to be extensions of each other visually and coherently, they have continuity. The motorcycle illustration in Fig. 11.3b gives a good example of continuity, where fuel tank, seat, side cover, and rear cowl appear as extensions of each other.
c. Feature lines: Feature lines are lines formed by visible boundaries of parts. The motorcycle in the example has the boundary lines of fuel tanks, the side cover and the rear cover having both continuity and similarity of the profile. So they impart a sense of unity to the design.
d. Character lines: Character lines are distinctly visible lines formed by the intersection between two surfaces. In the example illustration, all the three parts have the sharp straight lines forming the character lines which are sharp, straight, and parallel and so imparting unity to the design.
e. Graphics: Graphics, as we know are visuals (predominantly two-dimensional images) added to the visible surfaces. Unity of graphics means their mutual alignment, their alignment with character lines and alignment with feature lines.

Unity of design in a complex machine like a motorcycle can be derived by a combination of above parameters and unity between these parameters themselves. Table 11.1 shown here is an example in this direction. The table can be extended or shortened by adding or removing sub-factors, depending on the object under consideration, keeping, of course, the factors unchanged. For example, you can have motorcycles without visors or visors with windshield. Similarly, you can have motorcycles with more parts as part of front body or rear body as in the case of bigger motorcycles. Then, there are motorcycles with separate seats for rider and pillion, sometimes with partitions or backrests. In all such cases, the unity-calculation table will change. The basic point is, the table has to be exhaustive and should include all parts which build the visual character of the motorcycle. Based on this exhaustive table, the numerator and denominator of the unity factor

Table 11.1 Unity level calculation

Factor	Sub-factor	Marks	Explanation	Hunk score
Unity of form	Visor/fuel tank	0–1	0: Not similar or complimentary 1: Similar or complimentary	1
	Fuel tank/seat	0–1		1
	Fuel tank/side cover	0–1		0
	Side cover/side cowl	0–1		0
Continuity	Fuel tank/seat	0–3	0: No continuity 1: Just touching 2: End profile merging 3: Visually united	3
	Fuel tank/side cover	0–1	0: No continuity 1: Continuity	1
	Side cover/side cowl	0–1		1
Character lines		0–2	2: Continuity or alignment 1: No character lines 0: Dissonance	2
Feature lines	Visor/fuel tank	0–1	0: No alignment 1: Aligned	1
	Fuel tank/seat	0–1		1
	Fuel tank/side cover	0–1		1
	Seat/muffler	0–1		1
Graphics	Alignment	0–1	0: No alignment 1: Aligned	1
Character lines versus feature lines	Visor	0–1	0: No alignment 1: Aligned	1
	Fuel tank	0–1		1
	Side cover	0–1		1
Character lines versus graphics	Visor	0–1	0: No alignment 1: Aligned	Not applicable
	Fuel tank	0–1		1
	Side cover	0–1		1
Feature lines versus graphics	Visor	0–1	0: No alignment 1: Aligned	Not applicable
	Fuel tank	0–1		1
	Side cover	0–1		1

have to be calculated. The denominator is the sum of maximum marks assigned to each sub-factor (the total of third column considering the maximum marks). The numerator is the total of all marks scored against each sub-factor and the overall score is the percentage calculated by using these two quantities. If the overall score is less than 50%, the unity level is 1, if it is between 50 and 80%, the level is 2, and above 80%, level is 3.

To illustrate an example, the unity score calculation for a Motorcycle Hero Honda Hunk (shown in Fig. 11.5) has been sampled in Table 11.1. You can see that this bike has no graphics on the visor.

So, the denominator is calculated as 23 and the sum of Hunk scores that form the numerator is 21. Therefore the overall score is 91% (21/23 * 100 = 91), which means the unity level of Hunk is 3.

11.2 Dynamism

One important factor needs to be noted. Motorcycle is a dynamic object and its dynamism cannot be established merely on the basis of static visual factors. The experience that a rider gets while driving the motorcycle plays an important role in evaluating the dynamism inherent in a vehicle. But, these aspects get covered in the "context" factor, where the type of motorcycle covers the engine performance, acceleration, and the rider's posture. Therefore, in this section, we cover only the visual aspect of dynamism.

But this is not the only confusion. We have to deal with one more. In our discussion on motorcycle aesthetics, the word "*dynamism*" is used in two different contexts and therefore has two different connotations. The first is "*dynamism*" the *emotion*, which we have discussed in previous chapters. The second is "*dynamism*" the *design factor*, which we are discussing in this chapter. So "*dynamism*" the *design factor* is one of the factors generating the "*dynamism*" the *emotion* and *dynamism* the *emotion* depends on "*dynamism*" the *design factor*. This confusion could have been avoided by choosing some other word for one of the two contexts. But I deliberately chose to be with the confusion and rely on the intelligence of the reader to understand the context.

Visually dynamism can be defined as the element of visual tension created by the impression that the visual we are seeing is not going to remain static and something is going to change immediately. This tension is generated by three elements of the visual:

a. The divergence and alignment of lines: We will call it visual dynamism. We can illustrate this example by contrasting the image of a standing tiger with a galloping tiger. Whereas in the first case the four legs of the tiger as vertically parallel and the torso is horizontal, in the case of the galloping tiger the four legs are diverging and the torso is inclined upwards in the direction of the movement. This divergence immediately makes us feel the possibility of imminent change. Here also character lines, feature lines, and graphics play an important role. But, what matters more here is their orientation and alignment with each other.
b. The visual impression of the center of gravity: This impression is generated by the height and length ratio. The more is this ratio, the more is the perceived height of the center of gravity and thus the possibility of falling.
c. The visual imbalance: Any balanced visual gives us the sense of stability. Conversely, visual imbalance creates the sense of instability and hence generates the visual tension. We call this element "balance".

Fig. 11.4 **a** Virgin and the Child with Canon van der Paele (1436) by Jan Van Eyck, **b** Christ Healing the Sick, an etching by Rembrandt

Balance: A look at the two artworks gives an idea about the idea of balance. In the first work in Fig. 11.4a, the two figures on the left and right of Madonna in the center counterbalance each other due to their distance from the center, their color, and their size. You will note that the figure farther from center is taller and is deeper in color whereas the right figure, which is nearer to the center, is smaller and lighter in color. All this creates an overall balance. Similarly, in the second work in Fig. 11.4b the image of Christ is in center and the lighter images on the left are counterbalanced by the darker shades of the structures, even though the images are fewer.

Balance, in itself, is neither desirable nor not desirable and the need for it depends on the emotion the artist or the designer desires to generate. The emotion of dynamism logically depends on the lack of balance, which in an artwork may be generally undesirable but for automobiles may be desirable based on the visual direction. So, we derive some idea about the elements of visual balance and conclude that visual balance results from three major factors—(a) Visual weight and (b) Visual direction (c) Distance of elements from optical center.

a. Visual weight: It is the perceived weight of a visual element. It is a measure of how much anything in a visual attracts the eye of the viewer.
b. Visual direction: It is the perceived direction of the visual forces. It is the direction we perceive an element to be moving in if it was given a chance to move based on the forces acting on it.
c. Distance of elements from optical center: The discussion on the visual direction takes us to the visual or optical center which further takes us to symmetry. On the basis of symmetry, we can classify visual balance in two categories—(a) Symmetrical balance (2) Asymmetric balance. Since two terms are self-explanatory, we will not go into their definitions. What we can safely say (ignoring rare exceptions) is that all motorcycles are symmetrically balanced in top and front views along the centerline. So what actually matters is the asymmetric balance in the side view. To determine this asymmetric balance we

need to find the centerline for which we need to find the center. This takes us to a very important question as to what is the optical center.

Optical Center: Objects and elements balanced around a point which is the optical center of the image. Bradley [12] defines it as a point that attracts the viewer's eye unless it is pulled elsewhere. By default, the viewer's eye starts from the upper left of the image and travels towards the lower right, passing through the optical center. The center is the point of intersection between the path of this travel and the vertical geometric centerline (see Fig. 11.5).

Now for mathematical quantification of visual balance, let us consider the analogy with mechanical balance. For an object to be mechanically balanced, the sum of moments of various elements should be zero, i.e.,

$$\sum m \cdot r = 0$$

where

m mass of element
r distance of the element from the center

But in case of visual balance, the impact of an element on visual balance is inversely proportional to its distance from center. So in this case

$$\sum W/r = 0$$

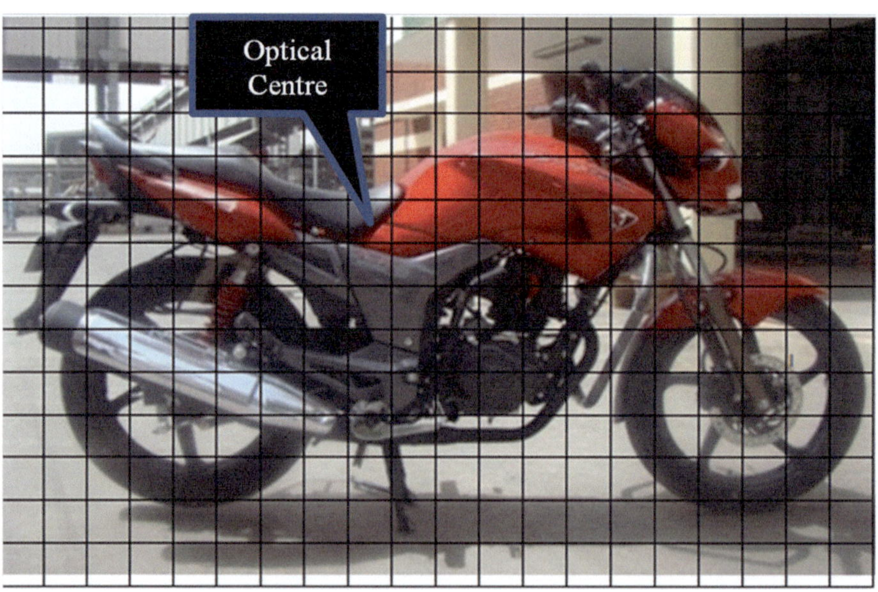

Fig. 11.5 Optical center and balance Grid

where

W total visual weight = wn

where

w visual weight of unit size,
n number of units,
r distance of the element from optical center.

Finally, we come to the basic question of quantification of the visual weight of a unit size. Bradley [12] lists the following factors impacting the visual weight:

a. Color: As we have discussed elsewhere, color has three elements—hue (location on color spectrum), value (percentage of whiteness) and chroma (absence of greyness). Steven assigns maximum weight to red and minimum to yellow and says that weight increases from cool to hot colors. Based on this we have the following scale (high to low from left to right in Fig. 11.6).
   Besides hue, the weight decreases with increasing value (darker color means more weight) and increases with chroma (deep color means more weight).
b. Value contrast: The same element has more weight when the difference of values between figure and ground is more.
c. Roughness: Visual perception increases the weight, whereas it reduces with transparency. On a scale of high to low from left to right, it can be represented as follows (Fig. 11.7).
d. Space: Positive space increases the weight and negative space decreases it.

Based on the above considerations, following scale can be generated for the weight of a unit (Table 11.2).

Then, roughness and space quotients are assigned based on their effect on visual weight (Table 11.3).

The horizontal and vertical imbalances are calculated by dividing the side view image of a motorcycle into grids of equal squares (see Fig. 11.5) and calculating the visual weights and imbalances using the following table. The various symbols are signified as follows:

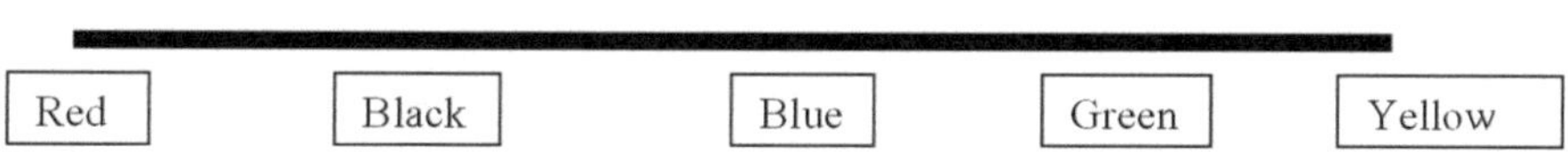

Fig. 11.6 Color spectrum for visual balance

Fig. 11.7 Roughness spectrum for visual balance

Table 11.2 Visual weight factor for colors

Red (rows: Chroma; columns: Value)

Chroma \ Value	1	2	3	4
1	2.25	2	1.75	1.5
2	2.5	2.25	2	1.75
3	2.75	2.5	2.25	2
4	3	2.75	2.5	2.25

Black

Chroma \ Value	1	2	3	4
1	2	1.75	1.5	1.25
2	2.25	2	1.75	1.5
3	2.5	2.25	2	1.75
4	2.75	2.5	2.25	2

Blue

Chroma \ Value	1	2	3	4
1	1.75	1.5	1.25	1
2	2	1.75	1.5	1.25
3	2.25	2	1.75	1.5
4	2.5	2.25	2	1.75

Green

Chroma \ Value	1	2	3	4
1	1.5	1.25	1	0.75
2	1.75	1.5	1.25	1
3	2	1.75	1.5	1.25
4	2.25	2	1.75	1.5

Yellow

Chroma \ Value	1	2	3	4
1	1.25	1	0.75	0.5
2	1.5	1.25	1	0.75
3	1.75	1.5	1.25	1
4	2	1.75	1.5	1.25

Table 11.3 Roughness and space quotients

Roughness quotient (add)

Rough	0
Glossy	-0.25
Plain	-0.5
Transparer	-0.75

Space quotient (Multiply)

High positive	1.5
Positive	1.25
Flat	1
Negative	0.75
High negative	0.5

N	No. of squares
r	Distance from the optical center in terms of no of squares
a	Color weightage based on hue, value and chroma (refer Table 5.4)
b	Value contrast factor
c	Surface texture factor
d	Space factor
w	Visual weight
W	Weighted visual weight
f	Multiplication factor based on no of squares as the value will be different if each square is considered separately. This factor removes this calculation error.

TVW Total visual weight on one side = $\sum W \cdot f = 0$

Horizontal Imbalance (%) = $(TVW_{front} - TVW_{rear})/(TVW_{front} + TVW_{rear})$
Vertical Imbalance (%) = $(TVW_{upper} - TVW_{lower})/(TVW_{upper} + TVW_{lower})$
An example has been shown for calculation of vertical imbalance in Hero Honda Hunk using the image and grids in Fig. 11.5 (Table 11.4).

Ratio: Length to width ratio of any object is an important visual indicator of its dynamism or stability and it has a significant engineering implication. The length to width ratio is a direct indicator of its center of gravity. The higher the center of gravity, the human perceives it to be more unstable and vice versa.

Perceived Center of Gravity: We perceive the dynamism in object through its tendency to change its position or location and the center of gravity is a direct indicator of this tendency. On one hand, beyond a point, the higher center of gravity of an automobile creates the perception of an acute case of instability, but within a point, it generates a sense of buoyancy and dynamism. Please look at the cars below. The second picture would have been taller than what it is; it would have created a sense of instability. But the current ration of length and height makes it more dynamic than the first one.

R_n factor: The most important aspect of a motorcycle's character is its dynamic behavior which can be characterized through the maneuverability–stability continuum. This continuum is predominantly defined by the geometry of the motorcycle as shown in Fig. 11.8a factoring the wheel base, wheel sizes, rake angle and the trail. The combination of these dimensions is one single mathematical entity which is designated by Cossalter [13] as "R_n factor" as shown in Fig. 11.8b. A low value of this factor indicates higher maneuverability meaning higher dynamism and a higher value indicates high stability.

Now, this Rn factor is not a visual factor. At the same time, the wheel base is the factor which creates the stability. So, the ratio L/H directly takes into account this sense of stability which is endorsed by dynamics as well.

Visual Dynamism: "The Ducati 900 is a finely engineered machine. My neighbors called it beautiful and admired its racing lines. The nasty little bugger looked like it was going 90 miles an hour when it was standing still in my garage" (Thompson 1995) [14].

No definition for visual dynamism can put it more succinctly. It is what generates a dynamic impulse with sheer visual power. Visual dynamism is characterized by the interplay of character lines, feature lines, and graphics.

Similar to the way we calculated the unity level of a design, the visual dynamism can be calculated using Table 11.5. A detailed example for calculation of overall dynamism has been shown in Table 11.6 which includes the calculation of visual dynamism as well.

Having defined dynamism, its binaries, its texts and sub-texts, ironically it still remains a psychological or emotional construct—at least in our perception. And therefore, we come across the vagaries of human perceptions. Quoting from the "Song of the Sausage Creature again" where Thompson compares the Ducati to a

Table 11.4 Table for calculation of balance

Model	Side	Details						w							$W = w \cdot n/r$	Multi factor (f)	$W \cdot f$	$\sum W \cdot f$	Imbalance
		Hue	Value	Chro	VC	Surf texture	Space	n	r	a	b	c	d	$w = (a + b - c)\, d$					
Hunk	Lower	Black	1	4	0	Rough	Flat	6	2.5	2.75	0.00	0.00	1.00	2.75	6.6	1.90	12.54	76.01875	14.86923
		Black	1	4	0	Plain	Flat	6	2.5	2.75	0.00	0.50	1.00	2.25	5.4	1.90	10.26		
		Black	1	4	0	Glossy	Flat	2	2	2.75	0.00	0.25	1.00	2.5	2.5	1.00	2.5		
		Red	3	4	0	Glossy	Flat	1	0.5	2.50	0.00	0.25	1.00	2.25	4.5	1.00	4.5		
		Black	1	4	0	Plain	Flat	2	1	2.75	0.00	0.50	1.00	2.25	4.5	1.33	5.985		
		Grey	1	1	0	Plain	Flat	6	1.5	2.00	0.00	0.50	1.00	1.5	6	1.50	9		
		Silver	4	3	0	Glossy	+ve	5.5	2.5	1.75	0.00	0.25	1.25	1.875	4.125	1.90	7.8375		
		Black	1	4	0	Rough	Flat	6	2.5	2.75	0.00	0.00	1.00	2.75	6.6	1.90	12.54		
		Black	1	4	0	Plain	Flat	3	2.5	2.75	0.00	0.50	1.00	2.25	2.7	1.00	2.7		
		Red	3	4	0	Glossy	+ve	0.5	1	2.50	0.00	0.25	1.25	2.8125	1.40625	1.00	1.40625		
		Black	1	4	0	Plain	Flat	0.5	0.5	2.75	0.00	0.50	1.00	2.25	2.25	1.00	2.25		
		Black	1	4	0	Plain	Flat	1	0.5	2.75	0.00	0.50	1.00	2.25	4.5	1.00	4.5		
	Upper	Black	1	4	0	Rough	Flat	4.5	2.5	2.75	0.00	0.00	1.00	2.75	4.95	1.00	4.95	56.33827	
		Red	3	4	0	Glossy	−ve	5	1.5	2.50	0.00	0.25	0.75	1.6875	5.625	1.50	8.4375		
		Red	3	4	0	Glossy	+ve	7	2	2.50	0.00	0.25	1.25	2.8125	9.84375	1.00	9.84375		
		Red	3	4	0	Glossy	+ve	3	2	2.50	0.00	0.25	1.25	2.8125	4.21875	1.00	4.21875		
		Grey	1	1	0	Plain	Flat	4	0.5	2.00	0.00	0.50	1.00	1.5	12	1.00	12		
		Black	1	4	0	Plain	Flat	1	0.5	2.75	0.00	0.50	1.00	2.25	4.5	1.00	4.5		
		Red	3	4	0	Glossy	+ve	2.5	3	2.50	0.00	0.25	1.25	2.8125	2.34375	1.00	2.34375		
		Black	1	4	0	Transparent	+ve	0.5	3	2.75	0.00	0.75	1.25	2.5	0.4166667	1.00	0.416667		
		Silver	4	3	0	Glossy	Flat	1.5	1.5	1.75	0.00	0.25	1.00	1.5	1.5	1.00	1.5		
		Black	1	4	0	Plain	Flat	2	3.5	2.75	0.00	0.50	1.00	2.25	1.2857143	1.00	1.285714		
		Grey	2	2	0	Plain	Flat	2	3.5	2.00	0.00	0.50	1.00	1.5	0.8571429	1.00	0.857143		
		Black	1	4	0	Plain	Flat	2	1	2.75	0.00	0.50	1.00	2.25	4.5	1.33	5.985		

back

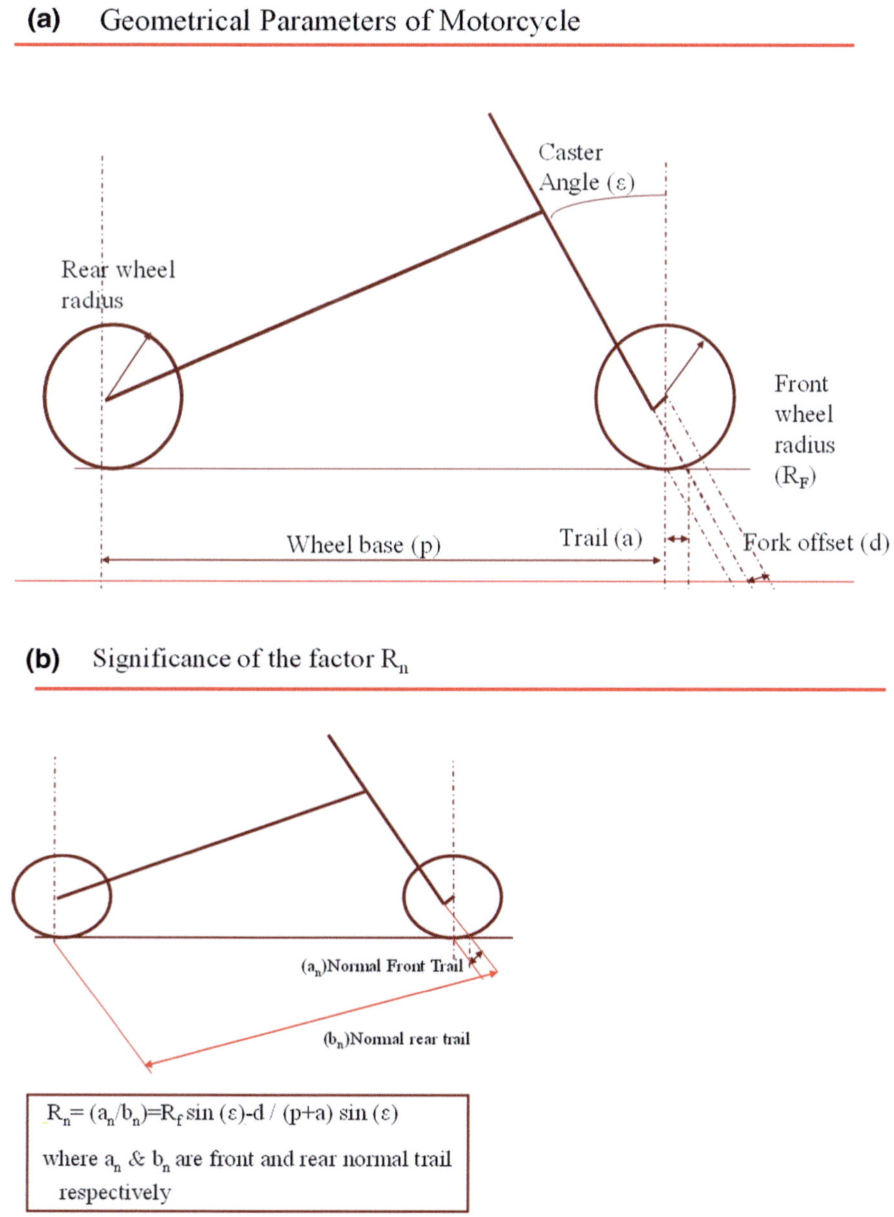

Fig. 11.8 **a** Basic geometry of a motorcycle, **b** The R_n factor explained

Table 11.5 Calculation of visual dynamism

Factor	Sub-factor	Marks	Explanation
Character lines	Visor	0–2	2: Diverging sharp 1: Diverging large fillet 0: Curve/straight/none
	Fuel tank	0–2	
	Side cover	0–2	
	Muffler	0–2	
Character lines alignment	Fuel tank/visor	0–1	1: Diverging and aligned 0: Straight or non-aligned
	Fuel tank/side cover	0–1	
Feature lines	Fuel tank	0–1	1: Inclined 0: Straight or curve
	Side cover	0–1	
	Seat	0–1	
	Muffler	0–1	
Character line and feature line matching		0–2	2: Fully matching(Diverging) 1: Partially matching (Diverging) 0: Not matching/straight
Graphics		0–2	2: Diverging horizontal 1: Diverging vertical 0: Straight or curve
Graphics and character lines matching		0–2	2: Fully matching(Diverging) 1: Partially matching (Diverging) 0: Not matching/straight
Graphics and feature lines matching		0–2	2: Fully matching(Diverging) 1: Partially matching (Diverging) 0: Not matching/straight

cannonball and feels it to be moving at 90 miles an hour, some others have found the same bike visually giving a backache.

Again for Thompson, Vincent Black Shadow (Fig. 11.9) was a bullet that went straight. Yet mathematically, it has neither visual imbalance or length to height ratio to match the perceptions. These vagaries of perception will always have exceptions. But ultimately, humans being rational animals, these exceptions will remain just that—exceptions.

So far, we have deconstructed the dynamism inherent in the design of a motorcycle into its elements—balance, ratio, and form. The point is to synthesize them into one single number which indicates the overall impact of all three. This magic number let us call it “overall dynamism” and designate by “D_o”. So let us describe this number by

$$D_o = (B_h - B_v)/100 + (1/r) + D_v$$

where,

B_h Horizontal imbalance
B_v Vertical imbalance
r Ratio
D_v Visual dynamism score

Table 11.6 Example for calculation of visual dynamism

Model	Character lines				Character lines alignment		Feature lines			
	visor	Fuel tank	Side cover	Mufflercover	Fuel tank/visor	Fuel tank/side cover	Fuel tank	Side cover	Seat	Muffler
	0–2	0–2	0–2	0–2	0–1	0–1	0–1	0–1	0–1	0–1
Hunk	2	2	2	2	1	1	1	1	1	1
Splendor+	0	0	0	0	0	0	0	0	0	0
CD-Deluxe	0	0	0	0	0	0	0	0	0	0
TVS Flame	2	2	2	2	1	0	0	0	1	1
TVS Star city	0	0	2	0	0	0	0	1	0	0
Passion pro	0	0	0	0	0	0	0	1	0	0
Splendor NXG	0	0	0	0	0	0	0	0	0	0
Glamour	1	1	1	0	1	1	0	1	0	0
Discover	1	1	1	0	0	1	0	1	0	0
CBZ	2	2	2	2	2	2	1	1	1	1
Pulser	2	2	2	2	1	1	0	1	1	1
Apache	2	1	2	2	0	0	0	1	1	1

Model	Character/feature line matching	Graphics	Graphics/character lines matching	Graphics/feature lines matching	Score	Visual score	L/H	HB	VB	Final score	Dynamism
	0–2	0–2	0–2	0–2							
Hunk	2	2	2	2	22	1.00	1.85	10.21	14.86	1.49	2.67
Splendor+	0	2	0	1	3	0.14	1.9	−4.63	16.9	0.45	0.80
CD-Deluxe	0	0	0	0	0	0.00	1.9	3.08	28.12	0.28	0.49
TVS Flame	1	2	1	1	16	0.73	1.85	−7.08	20.49	0.99	1.77
TVS Star city	0	2	0	0	5	0.23	1.85	2.68	6.75	0.73	1.30
Passion pro	0	2	0	1	4	0.18	1.9	1.74	0.79	0.72	1.28
Splendor NXG	0	2	0	1	3	0.14	1.9	-6.53	8.01	0.52	0.92

(continued)

Table 11.6 (continued)

Model	Character/feature line matching	Graphics	Graphics/character lines matching	Graphics/feature lines matching	Score	Visual score	L/H	HB	VB	Final score	Dynamism
Glamour	0	0	0	0	6	0.27	1.8	4.79	9.8	0.78	1.39
Discover	1	1	2	0	9	0.41	1.85	-0.22	3.84	0.91	1.62
CBZ	2	2	2	2	22	1.00	1.73	7.05	21.15	1.44	2.57
Pulser	1	2	2	2	20	0.91	1.79	23.59	2.4	1.68	3.00
Apache	1	0	1	0	12	0.55	1.74	17.8	12.34	1.17	2.10

Fig. 11.9 Vincent black shadow

Now, we calculate these score over a wider range with good sample size which includes all variations; these scores need to be normalized to scale of 3. This means the maximum score is given a normalized score of 3 and rest of the scores normalized proportionately. The example shown in Table 11.6 explains it all.

11.3 Unity and Dynamism

Coming back to the fundamental question how does the combination of unity and dynamism affects the emotions generated by an automobile (or, in this case, a motorcycle). In the end, it all boils down to the narrative, the motorcycle generates. We have seen it with fiction, poetry, and cinema. The narrative, of course, depends on the content. But we have seen in various movies based on the same content that the narrative is different and this difference is predominantly due to treatment. And what is this thing called "treatment"? Using the movie reviewer's jargon, it is the tightness of script ("unity" in design language) and flow ("dynamism" in design language), which define the narrative. So the emotions a movie generates depend on the interplay of unity and dynamism. Same holds true for poetry and fiction. And, of course, same holds true for motorcycle (or for that matter any artifact). So, it is this interplay of unity and dynamism which we have to establish, and establish quantitatively, to establish the relations between emotions and design.

The study conducted for this purpose involved collecting a set of three motorcycles for each combination. The exercise involved calculating the unity and dynamism using the methods illustrated above and placing the suitable combination

Table 11.7 Combination matrix for unity and dynamism

		Unity		
		Level 1	Level 2	Level 3
Dynamism	Level 1	 11	 12	 13
	Level 2	 21	 22	 23
	Level 3	 31	 32	 33

category. The matrix (Table 11.7) illustrates this (in this illustration, only one of the three bikes has been shown for each combination).

A group of around 300 respondents was asked to choose the dominant emotion for each of the motorcycle and intensity of the dominant emotion on a scale of 1–3 (it cannot be less than 1 as it will not remain the dominant emotion in that case). Then the results for each category were collected and plotted.

Figures 11.10, 11.11, 11.12 and 11.13 plot the variation of dynamism with unity for each emotion Sober, Glamor, Dynamism, and Intimidation.

X-axis represents unity score *Y* axis represents the intensity of emotion and so the diagram shows the interplay how the emotion varies with unity for a specific dynamism level. As the dynamism, level changes, the behavior of the emotion variation shifts dramatically. Let us see.

Sober: We observe a very clear trend that though sobriety is highest for dynamism level 1, it is highest at unity level 2. When unity level increases to 3, sobriety turns into glamor, which is illustrated in the diagram for glamor. Let us understand the implication on the ground. At low dynamism, you can note the sobriety in form of simplicity in motorcycle 11. As we move to motorcycle 12 with unity level 2, the sobering increases in form of classiness. This is the point of inflection. Beyond this point, classiness turns into glamor resulting into low sobriety, which can be noted in motorcycle 13. As the dynamism goes up to

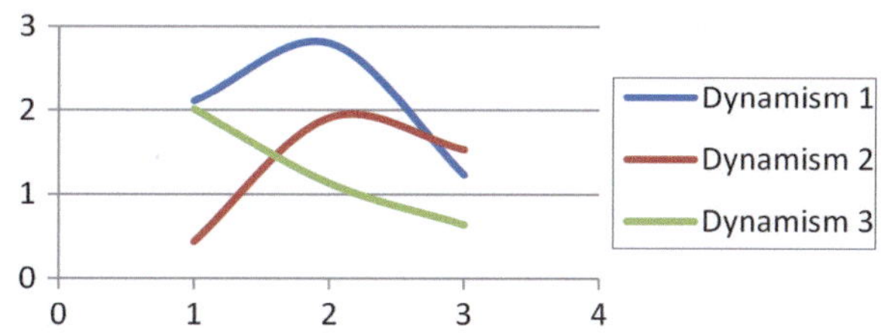

Fig. 11.10 Effect of unity and dynamism on "Sober" quotient

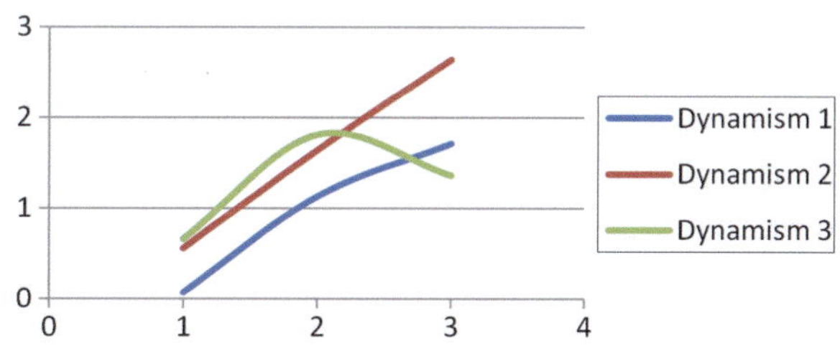

Fig. 11.11 Effect of unity and dynamism on "Glamor" quotient

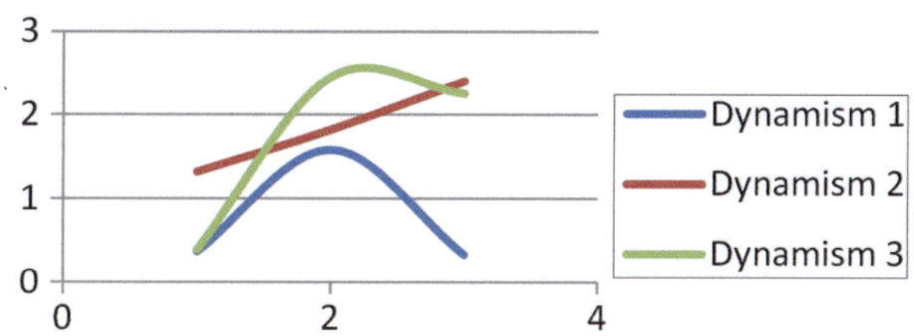

Fig. 11.12 Effect of unity and dynamism on "dynamism" quotient

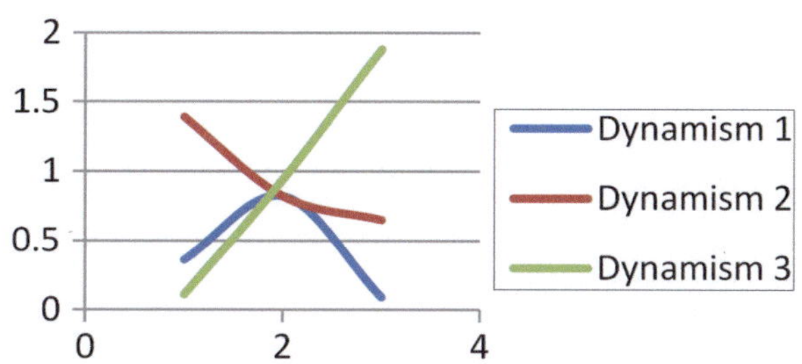

Fig. 11.13 Effect of unity and dynamism on "intimidation" quotient

medium, the pattern remains same but the quantum of sobriety is less, which is natural because of increasing dynamism, by its definition, brings down sobriety. Beyond unity 2, we again observe a change in behavior pattern. At high dynamism, the sobriety starts at a higher sobriety level and goes down rather sharply with increasing unity as we can see in motorcycles 31, 32, and 33. This is because the dynamic and intimidating behavior, which cannot coexist with sobriety, increases with increasing unity at high dynamism in design.

Glamorous: For high dynamism levels, glamor factor is highest at unity level 2, whereas for medium, it increases steadily with unity. If we increase the dynamism level to 3, we see an improvement in dynamism and intimidation factor as illustrated in next diagrams. Overall, at low unity, it is low and generally improves with unity except at very high dynamism. This is because at high dynamism, very high unity generates dynamic and intimidating peaks which cause the dip in glamor factor. You note this change as you move from motorcycle 32 to motorcycle 33.

Dynamic: This is the interesting territory. Ideally, dynamism as the emotion should directly relate to dynamism as the design factor. For medium dynamism, it is actually so. But for low and high dynamisms, we see strange behavior. At low dynamism, dynamic emotion is low, this is understandable. What happens is as you improve the unity at low dynamism, dynamic emotion also improves and as you further improve the unity, dynamic emotion goes down to turn into glamor. This is strange but true. High unity at low dynamism in real life is exemplified by bikes like Passion without having a single sharp point and they do generate the emotion of glamor. But high dynamism is even more interesting. At low unity, high dynamism generates low dynamic emotion. Again, this is strange but here lies the complexity of the tale. High dynamism is able to generate dynamic emotion, only if it is coupled with unity. Otherwise, it generates a good sober feeling only. Dynamic emotion improves with unity, but again at very high unity turns into intimidation. We can realize this looking at Suzuki Hayabusa which is shown as motorcycle 33.

Intimidating: Here we observe that highest level of intimidation is achieved at high dynamism, where the intensity is almost directly proportional to the unity level. At low dynamism level, it improves from low to medium whereas, at medium dynamism, it reduces with unity.

General trends observed in this survey are:

a. Lower levels of unity and dynamism generate a sober character.
b. Very high degree of unity with low dynamism generates glamorous character.
c. As we increase dynamism level with medium level of unity design moves towards higher dynamic character.
d. As we increase dynamism with high level of unity, design moves towards higher intimidating character.
e. The most interesting interplay of emotions is observed when intimidation and glamor are almost opposite to each other where dynamism and glamor are supplementing each other. Similarly, intimidation and dynamism are lowest at low dynamism where sobriety is highest, whereas sobriety and glamor supplement each other.

To me, this interplay of unity and dynamism is the key to aesthetics. This is the mysterious X factor, which the proponents of Gestalt theory believe is the difference between the whole and the sum of parts and decodes the mystery of the sub-text.

References

1. Pratt, Richard J. 2011. How Perception Works and its Role in Creating Engaging Design. *Design Principles and Practices International Conference*, Italy: Sapienza University of Rome.
2. Campbell, I. G. 1940. Factors which work towards Unity or Coherence in Visual Design. Elsevier. 146–162.
3. Kotnik, Toni. 2011 Algorithmic Design: Structuralism Reloaded, 327–335.
4. Durfeld, Michael. 2011. Ornamental Structuralism: From Rhythm to Evolution–the Future of Structuralism? 2011. 1–9. Available at http://duerfeld.de/Michael_Duerfeld_Ornamental_Structuralism_2011.pdf.
5. Cruickshank, Leon. 2011. The Case for a Re-evaluation of Deconstruction and Design; Against Derrida, Eisenman and their Choral Works. The Radical Designist. 1–9.
6. Young, Frank M. 2005. *Visual Studies, A foundation for artists and Designers*, 6–191 Eaglewood Cliffs, USA: Prentis Hall.
7. Elam, Kimberly. 2001. *Geometry of Design*, 6–98. USA: Princeton Architectural Press.
8. McClurg-Genevese, Joshua David. 2005. The Principles of Design. Available at http://digital-web.com/articles/principles_of_design.
9. Jirousek, Charlotte. 1995. Art Design and Visual Thinking. Available at http://char.txa.cornel.edu.
10. Lauer, David A., and Pentak, Stephen. 2007. *Design Basics*, 28–226. USA: Thomson Wadsworth.
11. Zelanski, Paul., and Fisher, Mary Pat. 1996. *Design Principles and Problems*, 34–35.USA: Wadsworth Cengage Learning.
12. Bradley, Steven. 2010. 19 Factors That Impact Compositional Balance. Vanceodesign. 1–16.
13. Cossalter, Vittore. 2010. *Motorcycle Dynamics*, 2: 57–59. USA: Race Dynamics, Greendale, WI.
14. Thompson, Hunter S. 1995. Song of the Sausage Creature. Cycle World (March,1995), Available at http://www.latexnet.org/~csmith/sausage.html

Chapter 12
Graphics: To Be or Not to Be

> It is intuitive that I can seduce a lady, a potential partner in an important business, or a corrupt politician, either by saying that I am rich and generous or by offering her or him a titillating dinner in the most luxurious restaurant of the city, with a menu that would have syntagmatically delighted Roland Barthes. It is equally intuitive that probably the dinner would be more convincing than a crude verbal statement. It is not intuitive why all this is intuitive. Perhaps it is by virtue of a "shallow" similarity in their effect that one intuitively understands that both behaviors produce ideas and emotions in the mind of the potential victim. But, in order to explain how both behaviors produce the same effect, one should look for something deeper. To look for such a deeper common structure, for the cognitive and cultural laws that rule both phenomena such is the endeavor of a general semiotics.
>
> —Umberto Eco in "Semiotics and the Philosophy of Language"

Semiotics is the science of signs and graphics in motorcycle are actually signs which serve complex functions of codes and metaphors to create a narrative, but at the end of the day as Eco says, achieve one simple objective—seduce the potential user by producing ideas and emotions in the mind of the potential user (I will not use the word "victim"). In business jargon, this is a marketing exercise called branding. But since it has a strong association with user's perception, it is an in-built characteristic of design. Manufacturers make it a point to convey messages regarding their value systems and product attributes through the design of the product itself. Generally, these messages are complex and are a mix of company values, product attributes, and character and the attributes of the target user. But above all, it tries to create a narrative, which can develop a sustainable relationship with the users. The key element of this narrative is to create congruence between the product, the producer, the message, and the messenger. To quote Eco again-sign is a link between the signifier (in this case, the bike) and signified (the desired message). Graphics play the role of a messenger who has to constantly emit coded messages about the product and the company to the user.

Graphics act at various levels. At one level, they have to act as the verbal and visual metaphor for the hedonistic benefits offered by the product, at another level it has to truthfully act as the teaser ad for the design attributes of the product. Considering the twin function of the graphics to inform (about the company and the

S. Chandra, *Aesthetics: Quantification and Deconstruction*,
Design Science and Innovation, https://doi.org/10.1007/978-981-10-6235-3_12

product) and advertise (metaphorically about the attributes) motorcycle graphics can be neatly divided into two parts—logo and the stripes. The logo itself has a huge metaphorical significance and is a mix of symbols and texts. On the other hand, stripes are pure artwork meant to act as the costume covering a beauty queen —to accentuate the attractive secrets of the object it is supposed to cover.

Logos are used to inform the user about the manufacturing company and the product name. But more than this purpose, they are important carriers of company and product's brand image. They are either three-dimensional objects (either plastic or metallic) or wrapped around a prominently visible parts. Logos essentially have three elements—company logo, company name, and product name. To take a simple example a Toyota car has a Toyota symbol, the text "TOYOTA" and the text for the name of the particular car, e.g., "Altis". The preferable qualities for a company logo are simplicity (generally companies prefer a logo which can be drawn with a simple pencil using a not so exceptional memory except for some remarkable exceptions where they go for deliberately complex ones); exclusivity, metaphorism, and high recall value. The design characteristics needed are usually high unity and dynamism especially for bikes. So, as far as the company name and product name are concerned, it is hardly the designer who decides them but the typography has to be decided by the designer and we will talk about typography in the later portion of this chapter. Important to note about all the three is that the designer has to decide their locations, their sizes, and whether they will be embedded with the stripe or will they be independent of the stripes.

Though graphics are not an engineering parameter in the strict definitive sense, it is an important contributor to the overall aesthetics of a motorcycle. Sometimes it has been observed that the same motorcycle with different graphics revolutionizes its aesthetic character. Motorcycle graphics have two important elements—logo and illustrations. Sometimes, they are separated and sometimes integrated.

12.1 Graphics

How do the graphics seduce us to create the emotions they create? In my view, this is the most complex working as the factors responsible are almost completely psychological and completely devoid of mechanical aspects. These psychological aspects, first and foremost depend on—(1) whether the logos are completely free from stripes, (2) whether the logos are embedded in the stripes, and (3) whether the stripes are completely independent of logos. These three cases have been studied separately.

Only logo: Logos are mostly found in two types of construction, i.e., metallic and stripes and each of them is applied either in flowing fonts or in separated fonts. On the other hand, they are either placed in a straight direction (vertically or horizontally) or placed along character lines. Another factor which affects the emotion they create is their color. They are available either in chrome, golden, dark

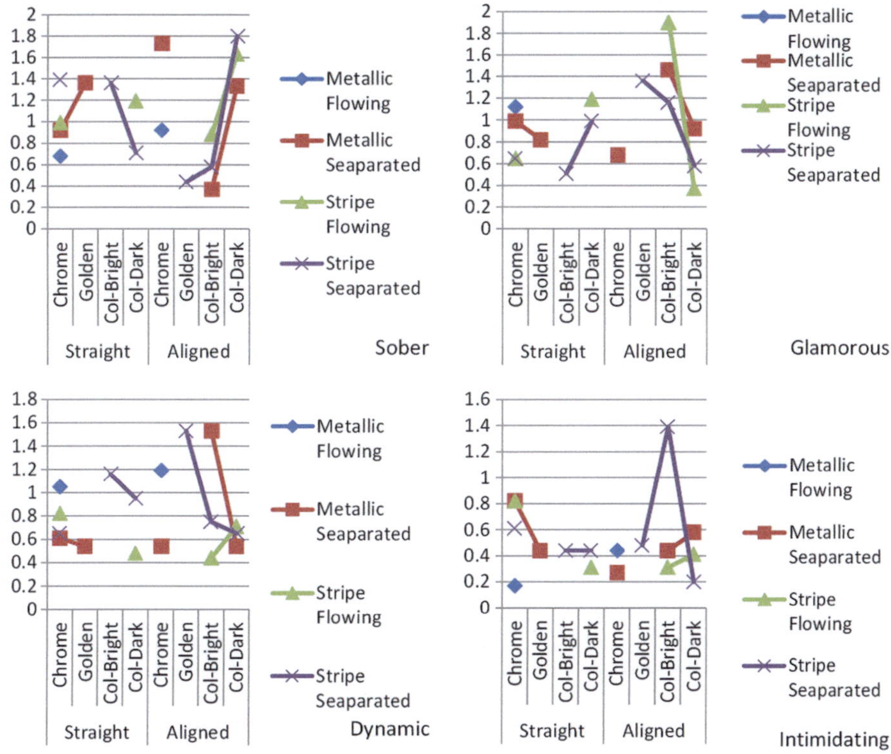

Fig. 12.1 The effect of logo on emotions

colors, or bright colors. With these broad classifications, their impact on the emotions generated has been illustrated in the figures (see Fig. 12.1).

When logo is deployed without an accompanying graphic (see Fig. 12.2), metallic graphics in separated fonts in chrome finish (Fig. 12.2a) and decals in separated fonts in dark colors (Fig. 12.2b) generate maximum sober emotions as long as they are aligned to the flow lines of the bike though the degree is not much to talk about. On the other hand, flowing stripes in bright colors (Fig. 12.2c) produce the best glamor appeal again, only when aligned to the character lines. Metallic and graphic brand names in separated fonts in aligned state (Figs. 12.2d, e) also produce a dynamic effect in bright and golden colors respectively. Intimidation is also generated by separated stripes in bright colors in aligned state (Fig. 12.2f) but the intensity of emotion is even lesser in case of dynamism and intimidation.

Another class of graphics is those where logos are embedded with stripes. Here also the categories are same as the case we have discussed earlier (see Fig. 12.3).

One thing that is clear at the outset is that the graphics increase their capability to generate emotions with more intensity when logos are embedded with stripes (see Fig. 12.4) compared to the case of logos alone. We can observe straightaway that the maximum intensity generated is beyond 2.5. But this observation is not valid

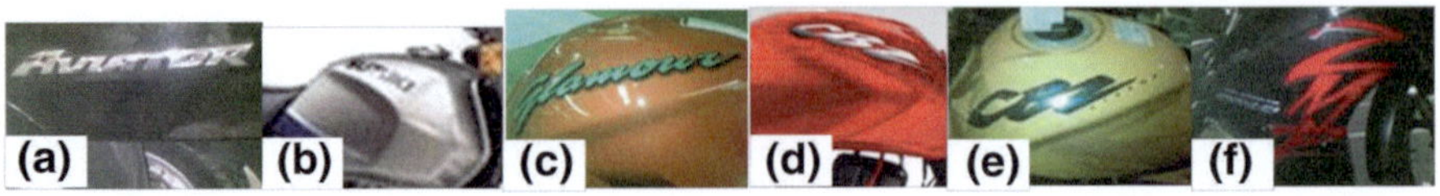

Fig. 12.2 Varieties of logos

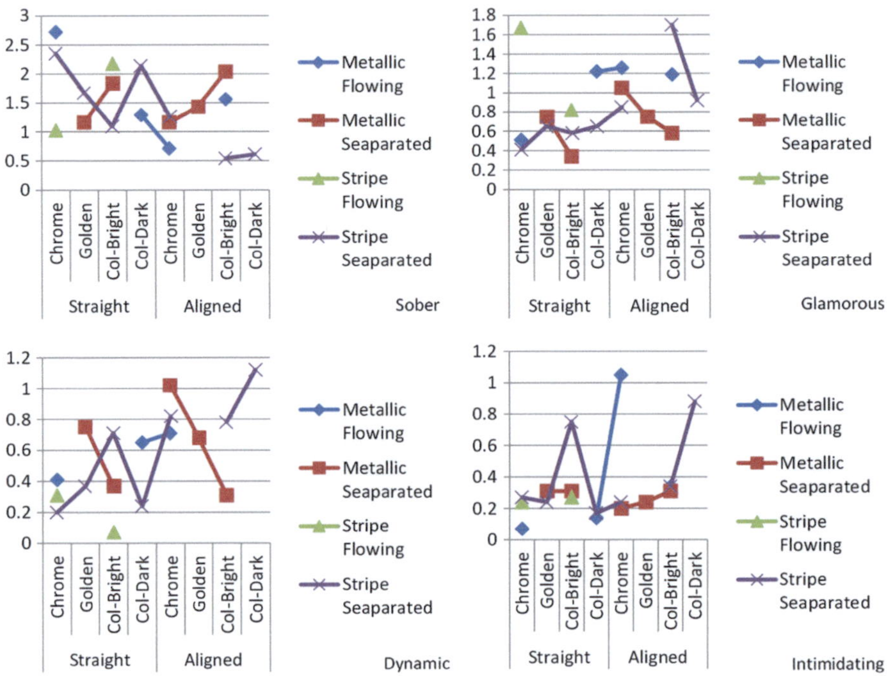

Fig. 12.3 The effect of logo with stripes

Fig. 12.4 Varieties of logos with stripes

across emotions and in fact, the maximum intensity is less in case of dynamic and intimidating emotions. You will also observe, almost across emotions that it is possible to produce good intensity with straight graphics as well. Talking about specific emotion, sober emotion is generated best with flowing metallic graphics (Fig. 12.4a) and separated stripes (Fig. 12.4b) come a close second. Please note that both these graphics are straight and are in chrome. Flowing stripes, straight again,

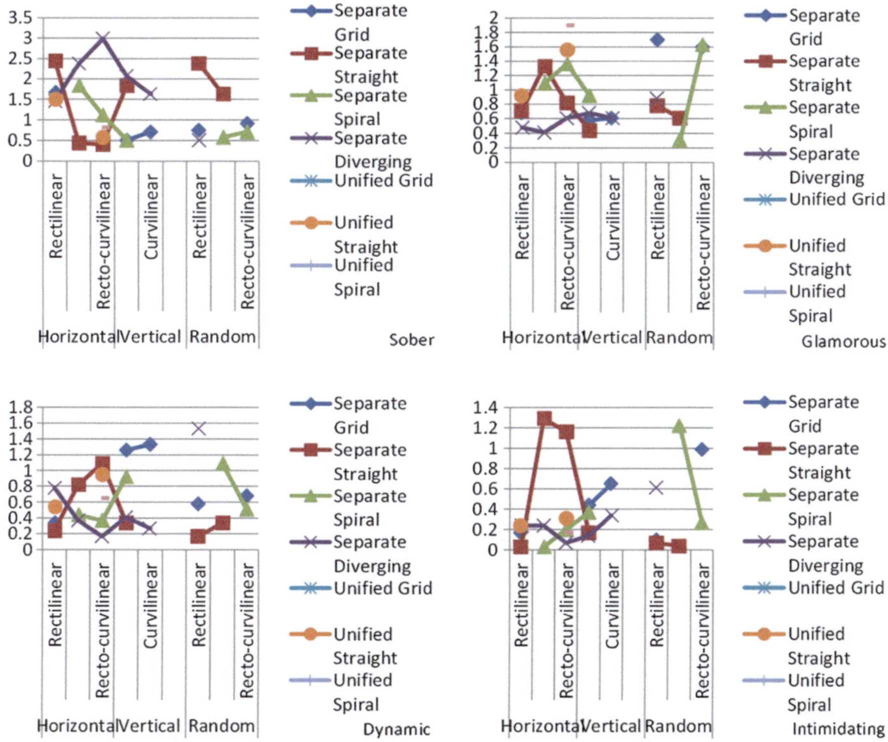

Fig. 12.5 The effect of independent stripes

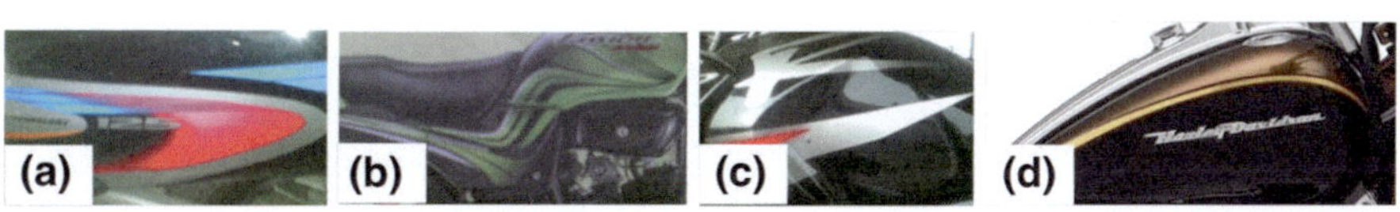

Fig. 12.6 Varieties of independent stripes

in chrome (Fig. 12.4c) produce the maximum glamor appeal. Dynamism comes best with aligned separated stripes in dark color (Fig. 12.4d) and intimidation with flowing metallic in chrome (Fig. 12.4e) and in aligned direction.

Now in case of independent graphics, logos and brands either do not accompany the stripes or have a separate entity. Since, here the stripes are free from the overpowering influence of logo, the total paradigm changes here. (see Fig. 12.5).

We can clearly see from these trends that graphics work best when they are freed from the influence of logos and intensity of emotion reaches its peak in this scenario (see Fig. 12.6). For sober stripes, recto-curvilinear separate spirals in the horizontal direction (Fig. 12.6a) appear to be generating most intensity though separate straight stripes do well in general. Unified stripes (Fig. 12.6b) are the preferred flavor for glamor. Random and vertical are the preferred directions for dynamism with separate diverging stripes (Fig. 12.6c) generating the best dynamism and separate straight style in the horizontal direction being the most intimidating (Fig. 12.6d).

Now, having seen through all classes of graphics, we can draw some conclusion across the classes. But before we come to this, some points need to be cleared at the outset. Logos simply cannot be wished away. So the choice we have is whether logo needs to be embedded in the graphics or to be designed as a separate entity. Another choice we have is whether to have graphics with logo or graphics are to be dispensed with altogether. A view across the classes will give us some clues.

Sober: Logos embedded in the design of graphics or graphics designed separately and free from the influence of logos are good for sober designs. The graphic styles most suitable are separate spiral or separate straight. The graphic designs of Splendor are the finest example where both cases can be seen.

Glamorous: Here, the issue needs to be seen in context, as the graphic design has to match with overall construction and form. For a normal three-box construction with overt positive–negative space combination with sharp fillets, only logos without stripes are the best with flowing design aligned along the character line. These are cases where the designer intends to have glamor and dynamism going together and creating the blend of both. This case is best exemplified by Hero Glamor. On the other hand, there are cases where glamor is generated by unified construction having no sharp lines and mild positive–negative space combination. These are cases where glamor is intended to fuse with a sense of serenity like Hero Passion. Unified graphics are most suitable for this design because this accentuates the unified character of the body.

Dynamic: This is a case where generally designers want to go without stripes altogether or minimalist, sharp, and diverging. In case designers intend to fuse dynamism with intimidation, minimalist recto-curvilinear stripes can be used as Harley generally does.

Intimidating: Intimidation is achieved best either without stripes and separated straight logo or separated straight stripes in horizontal direction.

12.2 Typography

The most important aspect of the logo is its typography (both for the model name and company name). Typography is a science developed through centuries and various typographical styles have evolved. The different stylistic aspects of typography, which decide the character of the bike, are as follows:

(a) The size
(b) The continuity or flow
(c) The sharpness of joints

Table 12.1 Hierarchy of typographical styles

Classical	Text	
	Roman	Old style
		Transitional
		Modern
Lineal	Sans Serif Square Serif	
Fancy	Ornamental Novelty	
Handwritten	Cursive Script	

(d) The number of geometric elements

a. The number of straight elements
b. The number of curved elements

All these elements with their various combinations are covered in typographical styles evolved through years. Sarkar [1] describes these styles to be divided into four groups (see Table 12.1):

Sarkar further describes the criterion for choosing a typeface as follows:

(a) Appropriateness
(b) Readability and legibility

The aspects which need to be decided for a typography design are as follows:

(a) Capital versus lower case
(b) Size
(c) Inclination

Effect of Case: The impact of these factors on their capacity to incite emotion has been illustrated in the following graphs (Fig. 12.7). In the first set of graphs, the effect case has been shown. *X*-axis shows the number of style, inclination, and size combination and *Y*-axis shows the intensity of emotion.

In quantitative terms, Table 12.2 shown here gives a more vivid picture. The mean values tell us a broader picture about which case is more effective in generating a specific emotional effect. The table showing maximum effect guides the designer to leverage an emotion to maximum effect. But the graphs above tell another story, which the tables do not tell. Though upper case is more effective for all four emotions, the story is not as simple as it looks. For sober, glamorous, and dynamic emotions, there are many combinations of style inclination and size when lower case is more effective. Designer has the freedom to use lower cases as well if they opt for these combinations.

Effect of size: Looking at Fig. 12.8, the trend is an open and shut case. Except for sober emotion, where both big and small fonts are equally effective, big size

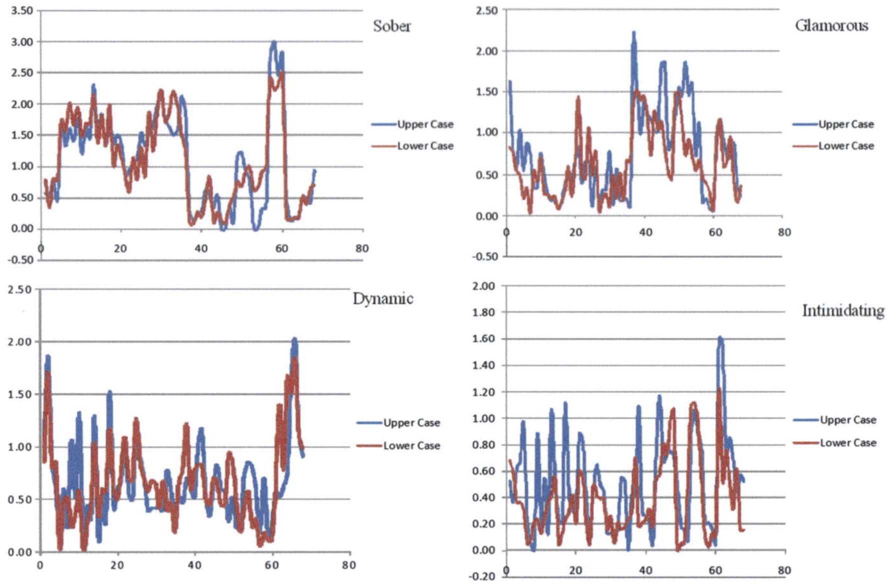

Fig. 12.7 Effect of case on emotion

Table 12.2 Effect of case on emotion ("Mean" and "Max" values)

Mean			Max		
	Upper	Lower		Upper	Lower
S	1.09	1.09	S	3.00	2.49
G	0.74	0.63	G	2.18	1.83
D	0.67	0.66	D	2.02	1.84
I	0.51	0.38	I	1.61	1.22

texts are effective for generating all other emotions barring some exceptional cases. The table for mean value and maximum value (Table 12.3) also validates the same conclusion.

a. Mean and max scores

The charts and table clearly indicate the bigger size gives a better result for all emotions except sober, whereas, for some individual styles, the smaller size can give better results. Finally, for highest results, the big size is definitely better.

Effect of inclination: The curves clearly indicate ambivalence (see Fig. 12.9). Except for dynamism, which shows a clear preference for inclined fonts, rest of emotions can be created by using either inclined or straight fonts. With mean and max score (see Table 12.4) there emerges a slightly more nuanced picture where straight fonts have a negligible edge over inclined ones for sober and intimidating emotions and inclined fonts have a similar negligible edge for glamor for mean score. But when it comes to creating a maximum possible impact, the haziness

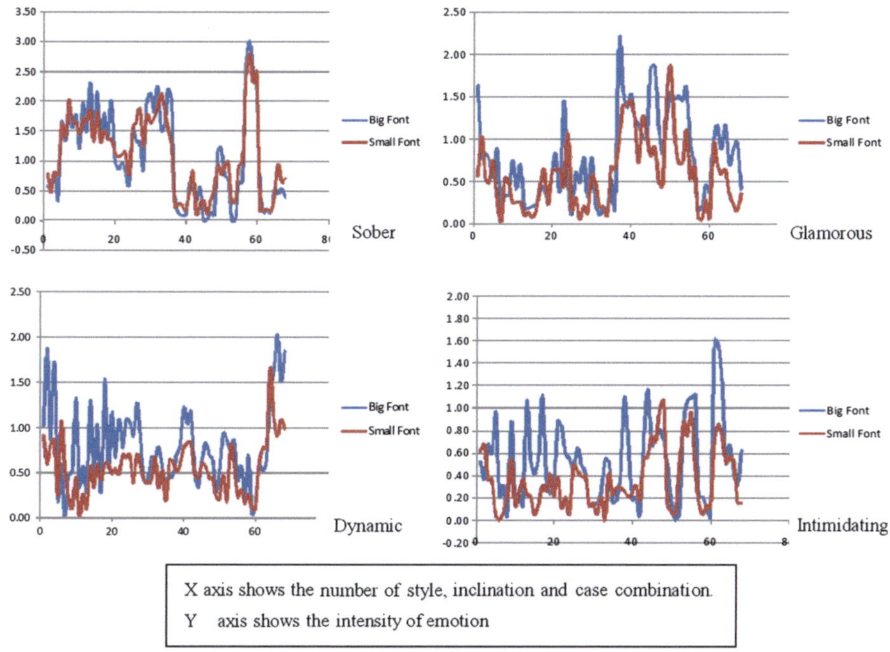

Fig. 12.8 Effect of typography size on emotion

Table 12.3 Effect of size on emotion ("Mean" and "Max" values)

Mean			Max		
	Big font	Small font		Big font	Small font
S	1.07	1.11	S	3.00	2.8
G	0.79	0.58	G	2.18	1.86
D	0.79	0.54	D	2.02	1.66
I	0.54	0.35	I	1.61	1.06

clearly vaporizes away. Inclined fonts have a clear and pronounced edge for sober and dynamic emotions whereas maximum glamor can be generated with inclined fonts. So designers have ample choices depending on the degree of emotion desired.

Max and mean scores:

So, not only the size, the case, the inclination, and the typography style is to be chosen very carefully by the designer to infuse the correct mix of emotions, he has to take the call whether to embed the text with the graphics or to keep them independent.

Finally, while designing the graphics, the designer has to think like a sculptor as he has to think of art in 3D, like a painter as he has to draw lines with color, like an engineer because he has to think of materials and their property, like a marketer to advertise the product and like sociologist to think of the unwritten social codes. Because at the end of the day, packaging does not sell a product but then it does help in selling it.

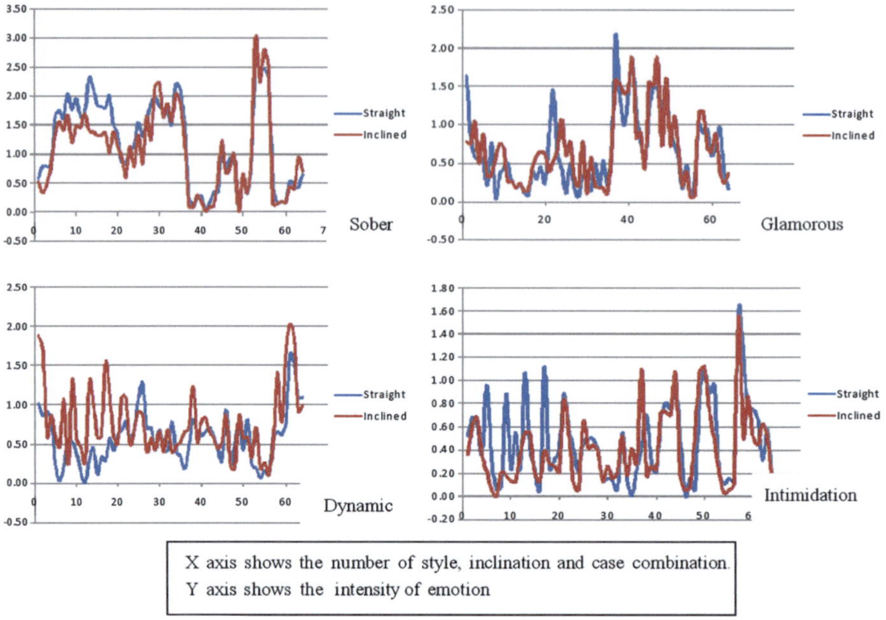

Fig. 12.9 Effect of inclination on emotion "Glamorous"

Table 12.4 Effect of inclination on emotion ("Mean" and "Max" values)

Mean			Max		
	Straight	Inclination		Straight	Inclination
S	1.09	1.06	S	2.46	3
G	0.64	0.68	G	2.18	1.86
D	0.56	0.76	D	1.63	2.02
I	0.48	0.41	I	1.61	1.55

Appendix 1: Graphics Combinations

Logo

		1	2	3	4	5	6	7	8
		Straight				Aligned			
		Chrome	Yellow	Coloured-Bright	Coloured-dark	Chrome	Yellow	Coloured-Bright	Coloured-dark
1	Flowing								
2 Metallic	Separated								
3	Flowing	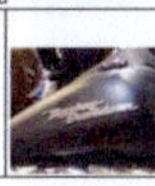							
4 Stripe	Separated								

Logo with stripe

		1	2	3	4	5	6	7	8
		Straight				Aligned			
		Chrome	Yellow	Coloured-Bright	Coloured-dark	Chrome	Yellow	Coloured-Bright	Coloured-dark
	Flowing								
Metallic	Capital								
	Flowing								
Stripe	Capital								

Appendix 2: Emotion Scores for Graphics Combinations

	S	G	D	I
11	0.68	1.12	1.05	0.17
15	0.92	0.68	1.19	0.44
21	0.92	0.99	0.61	0.82
22	1.36	0.82	0.54	0.44
25	1.73	0.68	0.54	0.27
27	0.37	1.46	1.53	0.44
28	1.33	0.92	0.54	0.58
31	0.99	0.65	0.82	0.82
34	1.19	1.19	0.48	0.31
37	0.88	1.90	0.44	0.31
38	1.63	0.37	0.71	0.41
41	1.39	0.65	0.65	0.61
43	1.36	0.51	1.16	0.44
44	0.71	0.99	0.95	0.44
46	0.44	1.36	1.53	0.48
47	0.58	1.16	0.75	1.39
48	1.80	0.58	0.65	0.2

(continued)

(continued)

	S	G	D	I
51	2.72	0.51	0.41	0.07
54	1.29	1.22	0.65	0.14
55	0.71	1.26	0.71	1.05
57	1.56	1.19	0.31	0.34
62	1.16	0.75	0.75	0.31
63	1.84	0.34	0.37	0.31
65	1.16	1.05	1.02	0.20
66	1.43	0.75	0.68	0.24
67	2.04	0.58	0.31	0.31
71	1.02	1.67	0.31	0.24
73	2. I8	0.82	0.07	0.27
81	2.35	0.41	0.20	0.27
82	1.67	0.65	0.37	0.24
85	1.26	0.85	0.82	0.24
87	0.54	1.70	0.78	0.34
88	0.61	0.92	1.12	0.88
91	1.67	0.68	0.34	0.17
94	0.51	0.58	1.26	0.44
95	0.71	0.61	1.33	0.65
97	0.75	1.70	0.58	0.10
99	0.92	1.16	0.68	0.27
101	2.45	0.71	0.24	0.03
102	0.44	1.33	0.82	1.29
103	0.41	0.82	1.09	1.16
104	1.84	0.44	0.34	0.17
107	2.38	0.78	0.17	0.07
108	1.63	0.61	0.34	0.34
112	1.84	1.09	0.44	0.03
113	1.12	1.36	0.37	0.20
114	0.51	0.92	0.92	0.37
118	0.58	0.31	1.09	1.22
119	0.71	1.63	0.51	0.27
121	1.46	0.41	0.78	0.24
122	2.38	0.48	0.37	0.24
123	2.99	0.61	0.17	0.07
124	2.07	0.68	0.41	0.14
125	1.63	0.61	0.27	0.34
127	0.51	0.88	1.53	0.61
141	1.50	0.92	0.54	0.24
143	0.58	1.56	0.95	0.31
163	0.82	1.90	0.65	0.17

Appendix 3: Emotion Scores for Typography Combinations

	1					2					3					4				
	Sample	S	G	D	I	Sample	S	G	D	I	Sample	S	G	D	I	Sample	S	G	D	I
A_1	ABCD	0.57	1.63	1.01	0.52	ABCD	0.52	0.78	1.86	0.36	Abcd	0.78	0.83	0.85	0.67	Abcd	0.34	1.75	1.71	0.60
A_2	ABCD	0.78	0.57	0.91	0.62	ABCD	0.47	1.04	0.60	0.67	Abcd	0.80	0.54	0.80	0.36	Abcd	0.78	0.49	0.85	0.36
B1	A BCD	0.26	1.74	1.24	0.08	A BCD	0.52	2.07	0.96	0.00										
B2	A BCD	1.06	1.19	0.49	0.13	A BCD	1.09	0.98	0.83	0.05										
C1	ABCD	1.63	0.54	0.21	0.96	ABCD	1.32	0.88	0.60	0.21	Abcd	1.74	0.21	0.03	0.31	Abcd	1.55	0.34	0.47	0.05
C2	ABCD	1.61	0.75	0.23	0.05	ABCD	1.40	0.34	1.06	0.00	Abcd	2.02	0.03	0.52	0.08	Abcd	1.66	0.54	0.23	0.21
D1	ABCD	1.76	0.34	0.52	0.88	ABCD	1.19	0.75	1.32	0.18	Abcd	1.94	0.41	0.36	0.23	Abcd	1.48	0.70	0.57	0.13
D2	ABCD	1.63	0.49	0.10	0.54	ABCD	1.45	0.26	0.44	0.13	Abcd	1.68	0.26	0.00	0.23	Abcd	1.68	0.26	0.26	0.36
E1	ABCD	2.30	0.18	0.31	1.06	ABCD	1.40	0.18	1.29	0.54	Abcd	2.15	0.21	0.44	0.41	Abcd	1.37	0.23	1.04	0.54
E2	ABCD	1.84	0.10	0.10	0.23	ABCD	1.32	0.13	0.57	0.21	Abcd	1.81	0.08	0.34	0.05	Abcd	1.31	0.16	0.60	0.13
F1	ABCD	1.79	0.39	0.28	1.11	ABCD	1.35	0.44	1.53	0.39	Abcd	1.99	0.28	0.57	0.23	Abcd	1.01	0.60	1.17	0.28
F2	ABCD	1.50	0.40	0.41	1.31	ABCD	1.37	0.65	1.60	0.26	Abcd	1.35	0.23	0.60	0.41	Abcd	1.14	0.62	0.49	0.21
G1	ABCD	0.85	0.83	0.67	0.88	ABCD	0.96	0.39	1.09	0.83	Abcd	0.83	1.45	0.78	0.60	Abcd	0.60	0.52	1.09	0.54
G2	ABCD	1.06	0.65	0.54	0.39	ABCD	1.11	0.65	0.49	0.10	Abcd	1.14	0.44	0.70	0.21	Abcd	0.78	1.06	0.67	0.05
H1	ABCD	1.53	0.10	1.06	0.47	ABCD	1.32	0.60	0.91	0.65	Abcd	1.29	0.49	1.27	0.49	Abcd	0.83	0.78	0.85	0.41
H2	ABCD	1.55	0.16	0.70	0.49	ABCD	1.63	0.31	0.39	0.44	Abcd	1.86	0.05	0.70	0.39	Abcd	1.24	0.21	0.59	0.39
I1	ABCD	1.97	0.31	0.41	0.13	ABCD	2.12	0.78	0.41	0.13	abcd	1.81	0.26	0.67	0.16	Abcd	2.23	0.10	0.65	0.26
I2	ABCD	1.76	0.13	0.41	0.13	ABCD	1.63	0.57	0.39	0.16	abcd	1.71	0.49	0.41	0.05	Abcd	1.86	0.21	0.67	0.21

	1					2					3					4				
	Sample	S	G	D	I	Sample	S	G	D	I	Sample	S	G	D	I	Sample	S	G	D	I
J1	ABCD	1.50	0.18	0.78	0.54	*ABCD*	1.55	0.18	0.57	0.52	abcd	2.20	0.52	0.39	0.16	*Abcd*	2.02	0.18	0.47	0.18
J2	ABCD	2.12	0.21	0.34	0.00	*ABCD*	1.89	0.10	0.52	0.41	abcd	1.50	0.67	0.18	0.23	*Abcd*	1.24	0.65	0.65	0.28
K1	ABCD	0.28	2.18	0.47	0.41	*ABCD*	0.10	1.58	0.70	1.09	abcd	0.08	1.37	0.80	0.70	*Abcd*	0.08	1.53	1.22	0.18
K2	ABCD	0.23	0.98	0.62	0.28	*ABCD*	0.28	1.40	0.52	0.26	abcd	0.26	1.40	0.60	0.21	*Abcd*	0.18	1.45	0.78	0.23
L1	ABCD	0.62	1.24	1.04	0.21	*ABCD*	0.57	1.17	1.17	0.05										
L2	ABCD	0.47	1.22	0.83	0.31	*ABCD*	0.83	0.73	0.83	0.18										
M1	A B CD	0.34	1.01	1.52	0.75	*A B CD*	0.54	1.01	0.34	1.17										
M2	A B CD	0.10	1.27	0.49	0.54	*A B CD*	0.26	1.04	0.44	0.57										
N1	ABCD	0.00	1.84	0.65	0.67	*ABCD*	0.00	1.86	0.83	0.73	Abcd	0.13	1.14	0.70	0.80	*Abcd*	0.08	0.83	0.65	0.73
N2	ABCD	0.34	0.80	0.60	0.75	*ABCD*	0.10	0.91	0.57	0.70	Abcd	0.34	0.52	0.44	0.98	*Abcd*	0.52	0.44	0.44	1.06
O1	ABCD	1.17	1.19	0.28	0.49	*ABCD*	1.22	1.55	0.52	0.18	Abcd	0.75	1.48	0.93	0.00	*Abcd*	0.67	1.50	0.85	0.05
O2	ABCD	0.93	1.45	1.23	0.16	*ABCD*	0.78	1.86	0.21	0.08	Abcd	0.91	1.09	0.47	0.05	*Abcd*	0.98	0.73	0.18	0.49
P1	ABCD	0.00	1.45	0.75	0.85	*ABCD*	0.00	1.61	0.85	1.06	Abcd	0.62	0.93	0.41	1.09	*Abcd*	0.65	0.73	0.57	1.11
P2	ABCD	0.31	0.73	0.80	0.88	*ABCD*	0.34	1.11	0.57	0.75	Abcd	0.91	0.54	0.23	0.96	*Abcd*	0.98	0.67	0.31	0.49
Q1	ABCD	2.69	0.16	0.18	0.21	*ABCD*	3.00	0.21	0.70	0.21	Abcd	2.41	0.47	0.05	0.10	*Abcd*	2.23	0.39	0.18	0.03
Q2	ABCD	2.46	0.08	0.16	0.16	*ABCD*	2.80	0.05	0.26	0.05	Abcd	0.44	0.98	0.57	1.61	*Abcd*	2.49	0.08	0.10	0.10
R1	A BCD	0.44	0.98	0.57	1.61	A BCD	0.13	1.17	0.52	1.55	A bcd	0.16	0.88	0.65	1.22	*A bcd*	0.13	1.17	1.40	0.52
R2	A BCD	0.16	0.93	0.62	0.73	A BCD	0.18	0.70	0.78	0.85	A bcd	0.18	0.60	0.78	0.85	*A bcd*	0.16	0.65	1.66	0.52
S1	A B CD	0.52	0.70	1.63	0.70	*A B CD*	0.44	0.88	2.02	0.44	A bcd	0.52	0.96	1.50	0.31	*A bcd*	0.39	0.41	1.84	0.62
S2	A B CD	0.41	0.36	1.09	0.57	*A B CD*	0.93	0.23	0.91	0.52	A bcd	0.65	0.16	1.09	0.16	*A bcd*	0.70	0.36	0.98	0.16

Reference

1. Sarkar, N.N. 1995. Art and production. Chapter 4 & 5, 34–65. Delhi: Sagar Publications.

Chapter 13
A Case Study

Till now, we have devised the methodology for quantifying and determining the emotions generated by the design of a motorcycle. We have also established the intensity of each emotion for each impacting combination and analyzed the general trends. But all this remains a hypothetical exercise unless we verify the validity of these observations. This can be done only by comparing the calculated values of the emotional quotients with actual observations. This whole exercise needs three major steps as follows:

a. Data collection for actual emotions generated by actual bikes
b. Evaluation of the same bikes for calculating the emotion quotients
c. Comparing the two values and reiteration of weightages for correct matching

Before we talk about the first step, let us talk about another story. We have talked about the story of motorcycle in India in another chapter. Now a question going around in my mind all the time, while I was conducting this study, was how do Indian motorcycles fare on the scale I have devised. So when I faced this issue of verifying my theory, I just lapped this opportunity up. Now, this step is not as simple as it looks on the surface because of the simplicity needed for this. Each step needs a judicious mix of practicality, feasibility, accuracy, and simplicity as a common observer cannot be expected to be very objective and quantitative about emotions and yet the results should reflect the correctness (or the absence of it) of our methodology. These steps for the data collection are:

a. Identification of vehicles
b. Collection of respondents' responses
c. Compilation of results

Now, the most important pre-requisite for this step is that it should reflect the true composition of Indian motorcycle market. So I chose all top selling bikes from all manufacturers in the popular segment except the niche bikes like Karizma from Hero and bikes from Royal Enfield as they cater to a very small segment and

S. Chandra, *Aesthetics: Quantification and Deconstruction*,
Design Science and Innovation, https://doi.org/10.1007/978-981-10-6235-3_13

observations of all respondents will generate an anomaly. (I conducted this study somewhere in 2012, so some of the popular bikes of recent vintage are missing. Anyway, the vintage of the bikes is not relevant for the main purpose so let us leave it aside.)

There is another issue, however. Or should I say, ulterior motive, to be honest? The coefficients that I have derived from the surveys need a validation. And for sure, when you validate them, some tweaking is practically inevitable. And so, be ready for some tweaking of the coefficients.

So, here is the list of bikes chosen (see Table 13.1).

Table 13.1 Motorcycles for physical survey

S. No	Motorcycle name	Manufacturer	Year of make	Photograph
1	Hunk	Hero Honda	2010	
2	CBZ Extreme	Hero Honda	2010	
3	Glamor	Hero Honda	2010	
4	Discover	Bajaj Auto	2010	
5	Apache	TVS	2010	
6	Pulser 150	Bajaj Auto	2010	

(continued)

Table 13.1 (continued)

S. No	Motorcycle name	Manufacturer	Year of make	Photograph
7	Star City	TVS	2010	
8	Flame	TVS	2010	
9	CD-Delux	Hero Honda	2010	
10	Splendor+	Hero Honda	2010	
11	Splendor-ISIXG	Hero Honda	2010	
12	Passion Plus	Hero Honda	2010	

This survey was conducted on the same respondents and the questionnaire was kept very simple. They had to simply identify the dominant emotion and its intensity on a scale of 1–3 (1 for low, 2 for medium, and 3 for high, well, it cannot be 0 if it is the dominant emotion), as the respondent cannot be expected to give relative rating for all emotions. Before we get down to the results, some mathematics to explain the calculation:

Total no of respondents = 235

For a particular motorcycle for a particular character,

No of respondents selecting that responding to that motorcycle = $\sum n$

Sum of scores for the character = $\sum x$

Score for the character $x = \sum x / \sum n$

Maximum score for any motorcycle for any character = $x_{\max}$
SGDI score for a particular motorcycle for a particular emotion $X = x/\, x_{\max} * 3$
See the compiled rated scores in Appendix 1.

13.1 Calculation of Evaluation Scores

You can see the calculation sheet for a bike for example in Appendix 2.

Before we move further, let us discuss how did we get there. Getting the score for form, context, and graphics is simple and explained in the calculation sheet itself. Then, for calculating the score for color we need to fix the texture score and here lies the problem. The bike in real life does not have a single value for texture but we need to arrive at a single value. So here is the solution.

Each bike was divided into equal area grids and the areas for each zone with different color and texture was calculated, and finally, the weighted SGDI score was calculated as shown in Table 13.2.

Texture scores are compiled in Appendices 3 and 4.

Next hurdle is the score for unity and dynamism. For calculation of dynamism for each bike the horizontal balance, vertical balance, ratio, and visual dynamism for each bike has to be calculated and then the final dynamism score can be calculated (Sample calculation has been illustrated in Chap. 11 on Unity and Dynamism).

Then unity can be calculated for each bike separately using the guidelines and then unity and dynamism score is easy.

13.2 Comparison and Iterations

So, now as we are ready with the rated and evaluated score for each bike, we can look at the comparison straightaway (Table 13.3). Here, the two emotions for each bike have been highlighted in green (for top) and blue (for next to top), both for physical evaluation and for calculated scores. You can very clearly visualize the matches and mismatches between the two types of scores.

Table 13.2 Calculation method for texture scores

Area	SGDI score				A*Score			
	S	G	D	I	S	G	D	I
A_1	s_1	g_1	d_1	i_1	$A_1 s_1$	$A_1 g_1$	$A_1 d_1$	$A_1 i_1$
A_2	s_2	g_2	d_2	i_2	$A_2 s_2$	$A_2 g_2$	$A_2 d_2$	$A_2 \mathrm{i}_2$
A_n	s_n	g_n	d_n	i_n	$A_n s_n$	$A_n g_n$	$A_n d_n$	$A_n i_n$
Final score					$\sum A_x s_x / \sum A$	$\sum A_x g_x / \sum A$	$\sum A_x S_x / \sum A$	$\sum A_x S_x / \sum A$

Table 13.3 Comparison of physical survey and evaluation results

Motorcycle	Physical Evaluation	Calculated Score
CD-DLX	2.75	2.56
	0.06	1.43
	0.27	2.10
	0.00	1.27
GLAMOUR	0.21	1.49
	2.66	2.61
	0.27	1.76
	0.00	1.03
PASSION	0.39	1.63
	2.18	2.81
	0.54	1.90
	0.00	1.09
SPLENDOR+	2.99	3.00
	0.09	1.49
	0.15	1.99
	0.09	1.31
HUNK	0.06	0.95
	0.21	1.81
	2.18	2.77
	1.12	1.69
CBZ EXTREME	0.00	0.98
	0.18	1.78
	2.06	2.68
	1.15	2.15
DISCOVER	0.85	2.07
	0.57	1.88
	0.97	2.31
	0.03	1.43
PULSER 150	0.15	1.66
	0.30	1.96
	1.85	2.68
	0.94	1.34
TVS APACHE	0.21	1.75
	0.27	1.96
	1.91	2.50
	0.54	1.43
TVS FLAME	0.48	1.75
	0.76	1.91
	1.06	2.61
	0.33	1.39
TVS STAR CITY	1.88	2.22
	0.21	1.85
	0.18	1.80

Important observations from this comparison are as follows:

i. The most dominant emotion exactly matched.
ii. The second most dominant matched in all cases except Hunk, Pulsar and Apache.
iii. There was mismatch in scores, which is expected.

Since the second dominant emotion was not matching in three cases, the scores for emotion factors need a slight tweaking. On closer examination two factors are observed as follows:

1. The score for intimidation for cast wheels with big wider tire (B2) is too low (i.e., 0.24) in comparison to score for glamor, which lowers the intimidation factors for Hunk. As a result, Glamor becomes the second dominant emotion

instead of intimidation. To correct this anomaly the scores for these two emotions needs to be adjusted.
2. The scores for glamor and intimidation for U2D3 and U3D3 are skewed in favor of glamor. This needs to be realistically adjusted.

With these observations, the following changes were made:

a. Wheels and tires (Table 13.4)
b. Unity and Dynamism (Table 13.5)

With these changes, the score charts for wheels and tires and Unity and dynamism are modified and the motorcycle evaluation charts are modified for Hunk, CBZ, Apache, Flame, and Pulsar and the comparison chart between these rated scores and physical evaluation score are prepared afresh (Table 13.6). Here also green and blue colors mean the top and next to top scores.

Now, we can observe that the most dominant and second dominant emotions for all bikes are exactly matching. The second point to be observed is the relative identification of the bikes. Here, for each emotion, we identify three most dominant bikes and see whether they match for physical evaluation and rated scores. This can be seen in Table 13.7.

Here we observe that for emotions S and G, all three bikes are matching (except that the fact the order is reversed in case of G for the top two between Passion and Glamor). For D and I, top two bikes are exactly matching.

The next criterion for evaluating the results is the relative position of models for each emotion. We plot the curves for physical evaluation and rated scores for each emotion. Here scores have been plotted for all bikes for each emotion both for evaluated and rated scores (Fig. 13.1). Let us observe the two curves for their variation patterns across the bikes. We observe that though the curves are shifted, the variation patterns are absolutely same.

The most critical observation that we made during this verification was that the calculated scores were quite close to the actual scores given by respondents, except for some. Still, we made some iteration to our SGDI scores for factors wheels and tires and unity and dynamism. After these iterations, we observe that the relative variation for each emotion is matching for physical evaluation and rated scores. The

Table 13.4 Correction in SGDI values for wheel and tires

	S	*G*	*D*	*I*
Before	0.32	1.9	0.93	0.24
After	0.32	1.4	0.93	0.72

Table 13.5 Correction in SGDI values for unity and dynamism

		S	*G*	*D*	*I*
U2D3	Before	1.13	1.81	2.45	0.94
	After	1.13	1.1	2.45	1.75
U2D3	Before	0.64	1.36	2.25	1.88
	After	0.64	1.05	2.25	2.08

Table 13.6 Comparison of physical survey and evaluation results

Motorcycle	Physical Evaluation	Calculated Score
CD-DLX	2.75	2.56
	0.06	1.43
	0.27	2.10
	0.00	1.27
GLAMOUR	0.21	1.49
	2.66	2.61
	0.27	1.76
	0.00	1.03
PASSION	0.39	1.63
	2.18	2.81
	0.54	1.90
	0.00	1.09
SPLENDOR+	2.99	3.00
	0.09	1.49
	0.15	1.99
	0.09	1.31
HUNK	0.06	0.95
	0.21	1.68
	2.18	2.93
	1.12	1.92
Motorcycle	Physical Evaluation	Calculated Score
CBZ EXTREME	0.00	0.98
	0.18	1.65
	2.06	2.81
	1.15	2.37
DISCOVER	0.85	2.07
	0.57	1.88
	0.97	2.31
	0.03	1.43
PULSER 150	0.15	1.32
	0.30	1.72
	1.85	2.94
	0.94	1.74
TVS APACHE	0.21	1.41
	0.27	1.72
	1.91	2.77
	0.54	1.82
TVS FLAME	0.48	1.75
	0.76	1.91
	1.06	2.61
	0.33	1.39
TVS STAR CITY	1.88	2.22
	0.21	1.85
	0.18	1.80

Table 13.7 Comparison of physical survey and evaluation results (relative position in market)

	Physical evaluation	Calculated score
S	Splendor	Splendor
	CD Dlx	CD Dlx
	TVS Star City	TVS Star City
G	Glamor	Passion
	Passion	Glamor
	TVS Flame	Flame
D	Hunk	Hunk
	CBZ Extreme	CBZ Extreme
	Apache	Pulsar 151
I	CBZ Extreme	CBZ Extreme
	Hunk	Hunk
	Pulser	Apache

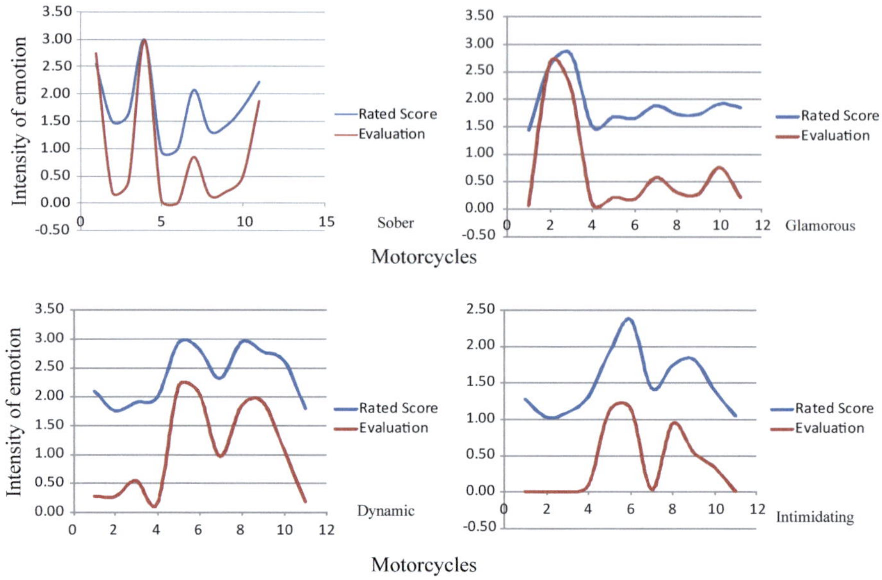

Fig. 13.1 Comparison of rated score and physical evaluation results

standard deviation for physical evaluation is slightly higher compared to the calculated ratings. This is, of course on expected lines as in our perception the differentiations have to be sharper to make them perceptible, whereas calculations are more objective and accurate.

Fundamentally, I wanted to conduct this study at two levels. On one hand, the purpose was to test the validity of the methodology established in this book. On the other hand, I wanted to understand the deeper psychological and social nuances of Indian motorcycle user. So, as far as the first objective is concerned, this study concludes that emotions generated by a design are scientifically identifiable and quantifiable. The mathematical tool generated is able to identify correctly the top two dominant emotions in all 11 motorcycles. The relative positioning in Indian motorcycle market is correctly identified to a great extent. For emotions *S* and *G*, the top three bikes are matching. For *D* and *I*, top two bikes are exactly matching. The variation pattern for all emotions also matches whereas the standard deviation for physical evaluation is slightly higher compared to the calculated ratings. This is understandable because in human perception, only the dominant emotions prevail and non-dominant ones score lower than their actual rating.

The larger fundamental issue here is that of deconstruction. The basic question, where we started was whether the aspect of emotions generated by a design is mathematically deconstructible. To a great extent, this question has been answered in affirmative. The aspect of the whole meaning more than the sum of parts is resolved with a quantification of unity and dynamism. Moreover, the parameter of context brings to the table the unstated sociocultural narratives.

Developing an understanding of the Indian motorcycle user is a real take-away from this study. Fundamentally, we can classify the user into three classes. At the starting rung of the ladder is entry segment consisting of motorcycles like CD-Deluxe, Splendor, Discover. The important discovery for a designer of this segment is the emotional need of the user which puts sobriety at the top and dynamism as the second important emotion. This is reflected in minimalist lean construction with no-nonsense attitude and without clutter. The second class is the user which puts glamor at the top and dynamism at the second desirable step. This class consists of bikes like Glamor and Passion where performance in form of acceleration is important but what matters most is glamorous looks reflected in unified construction and positive–negative space combination. The third category is the performance oriented user for whom dynamism followed by intimidation are the most desirable characteristics. The most important aspect of this classification is that the emotion profile of a bike generated by the visual design is a promise which must be fulfilled by the engineering performance. Otherwise, this dissonance can prove fatal.

Now, each society has its own social hierarchy and the segmentation of motorcycles depends on this social hierarchy, its power dynamics, and its aesthetic sensibility. For each market, this case study can be conducted after establishing the emotion scores and weightages to determine this segmentation. Moreover, these segmentations are not static. New products are introduced, new categories are introduced, the sensibilities of a society change and so change the segmentations.

But the fundamental purpose of this case study was not to determine the segmentations in Indian motorcycle but to test the validity of the methodology established which has been well achieved.

Appendix 1: Calculation of Evaluation Scores

Motorcycle	Character	$\sum x$	n	$x = \sum x/n$	Weighted evaluation survey score
CD DLX	S	*455*	210	1.94	2.75
	G	*10*	5	0.04	0.06
	D	*45*	20	0.19	0.27
	I	*0*	0	0.00	0.00
GLAMOR	S	*35*	20	0.15	0.21
	G	*440*	185	1.87	2.66
	D	*45*	25	0.19	0.27
	I	*0*	0	0.00	0.00
PASSION	S	*65*	30	0.28	0.39
	G	*360*	165	1.53	2.18
	D	*90*	40	0.38	0.54

(continued)

(continued)

Motorcycle	Character	$\sum x$	n	$x = \sum x/n$	Weighted evaluation survey score
	I	*0*	0	0.00	0.00
SPLENDOR+	S	*495*	215	2.11	2.99
	G	*15*	5	0.06	0.09
	D	*25*	10	0.11	0.15
	I	*15*	5	0.06	0.09
HUNK	S	*10*	5	0.04	0.06
	G	*35*	15	0.15	0.21
	D	*360*	135	1.53	2.18
	I	*185*	80	0.79	1.12
CBZ EXTREME	S	*0*	0	0.00	0.00
	G	*30*	15	0.13	0.18
	D	*340*	140	1.45	2.06
	I	*190*	75	0.81	1.15
DISCOVER	S	*140*	85	0.60	0.85
	G	*95*	65	0.40	0.57
	D	*160*	85	0.68	0.97
	I	*5*	5	0.02	0.03
PULSER 150	S	*25*	10	0.11	0.15
	G	*50*	20	0.21	0.30
	D	*305*	135	1.30	1.85
	I	*155*	70	0.66	0.94
TVS APACHE	S	*35*	25	0.15	0.21
	G	*45*	30	0.19	0.27
	D	*315*	140	1.34	1.91
	I	*90*	40	0.38	0.54
TVS FLAME	S	*80*	50	0.34	0.48
	G	*125*	60	0.53	0.76
	D	*175*	100	0.74	1.06
	I	*55*	25	0.23	0.33
TVS STAR CITY	S	*310*	200	1.32	1.88
	G	*35*	15	0.15	0.21
	D	*30*	20	0.13	0.18
	I	*0*	0	0.00	0.00

Appendix 2: Sample Calculation of Emotion Scores

	Weightage	S	G	D	I	Details
a. Color						
Body color	0.33	0	1.33	2.24	2.45	bl
Body/seat comb	0.33	0	1.33	2.24	2.45	bl
Body/stripe comb	0.33	0.1	1.24	1.45	1.86	bl/r2
Mean (a)	0.5	0.033333	1.3	1.976667	2.253333	
Texture (b)	0.5	1.589014	2.589014	3.589014	4.589014	
Weighted mean		0.811174	1.944507	2.78284	3.421174	
b. Form						
Fuel tank	0.31	0.79	0.68	1.68	0.11	MO3
Seat	0.34	2.08	0.08	0	0.12	
Visor	0.31	0.57	1.33	1.61	0.32	62
Side cover	0.04	0.64	1.81	1.32	0.25	3D
Weighted mean		1.1544	0.7227	1.0727	0.6033	
c. Graphics						
Typography	0.75	0.805	0.465	0.855	0.6	N14/H11
Graphics	0.25	2.45	0.71	0.24	0.03	
Weighted mean		1.21625	0.52625	0.70125	0.5202	
d. Context						
Type	0.54	3	0.09	0.14	0.23	
Fr facia	0.23	0.04	1.93	1.14	0.11	
Tires and wheels	0.23	0.81	1.58	0.85	0.12	
Weighted mean		1.8155	0.8559	1.4178	0.1771	
e. Unity and dynamism						
		2.8	1.13	1.58	0.82	U2D1
f. Overall						
Color	0.12	0.811174	1.944507	2.78284	3.421174	
Form	0.18	1.1544	0.7227	1.0727	0.6033	
Graphics	0.15	1.21625	0.52625	0.70125	0.5202	
Context	0.25	1.8155	0.8559	1.4178	0.1771	
Unity and dynamism	0.3	2.8	1.13	1.58	0.82	
Weighted mean		1.781445	0.995339	1.460664	0.88744	

Appendix 3: Sample Calculation of Emotion Scores for Texture

Model	Side	Details			Texture Score (x)				n	xn				Total score = ∑xn/∑n			
		Hue		Surf Texture	S	G	D	I		S	G	D	I	S	G	D	I
Hunk	LHS	Red	MH	Glossy	0.6	1.49	0.48	0	6	3.6	8.94	2.88	0	0.43	1.08	0.89	0.95
		Red	MH	Glossy	0.6	1.49	0.48	0	6	3.6	8.94	2.88	0				
		Silver			1.15	0.59	1.27	0.24	3	3.45	1.77	3.81	0.72				
		Black		Rough	0.3	0.45	0.36	1.49	6	1.8	2.7	2.16	8.94				
		Black		Plain	0.3	0.51	0.89	1.1	0.5	0.15	0.255	0.445	0.55				
		Red	MH	Glossy	0.6	1.49	0.48	0	2.5	1.5	3.725	1.2	0				
		Red	MH	Glossy	0.6	1.49	0.48	0	2.5	1.5	3.725	1.2	0				
		Chrome		Glossy	2.02	2.8	3	2.78	1	2.02	2.8	3	2.78				
		Chrome		Glossy	2.02	2.8	3	2.78	1	2.02	2.8	3	2.78				
		Black		Glossy	0.71	0.39	0.6	0.3	2	1.42	0.78	1.2	0.6				
		Black		Rough	0.3	0.45	0.36	1.49	7	2.1	3.15	2.52	10.43				
		Silver			1.15	0.59	1.27	0.24	3	3.45	1.77	3.81	0.72				
		White			1.65	0.41	0.12	0.22	1	1.65	0.41	0.12	0.22				
	RHS	Black		Rough	0.3	0.45	0.36	1.49	4	1.2	1.8	1.44	5.96				
		Black		Rough	0.3	0.45	0.36	1.49	8	2.4	3.6	2.88	11.92				
		Silver			1.15	0.59	1.27	0.24	4	4.6	2.36	5.08	0.96				
		Red	MH	Glossy	0.6	1.49	0.48	0	4	2.4	5.96	1.92	0				
		Chrome			2.02	2.8	3	2.78	7	14.14	19.6	21	19.46				
		Red	MH	Glossy	0.6	1.49	0.48	0	0.5	0.3	0.745	0.24	0				
		Silver			1.15	0.59	1.27	0.24	1	1.15	0.59	1.27	0.24				
		Black		Plain	0.3	0.51	0.89	1.1	2	0.6	1.02	1.78	2.2				

(continued)

(continued)

Model	Side	Details			Texture Score (x)				n	xn				Total score = $\sum xn/\sum n$			
		Hue		Surf Texture	S	G	D	I		S	G	D	I	S	G	D	I
TVS Star	LHS	Red	MH	Glossy	0.6	1.49	0.48	0	3.5	2.1	5.215	1.68	0	0.87	0.64	0.78	0.45
		Yellow	ML	Glossy	1.13	0.21	0.86	0.06	4	4.52	0.84	3.44	0.24				
		White			1.65	0.41	0.12	0.22	1	1.65	0.41	0.12	0.22				
		Black		Glossy	0.71	0.39	0.6	0.3	1	0.71	0.39	0.6	0.3				
		Silver			1.15	0.59	1.27	0.24	3	3.45	1.77	3.81	0.72				
		Silver			1.15	0.59	1.27	0.24	2	2.3	1.18	2.54	0.48				
		Red	MH	Glossy	0.6	1.49	0.48	0	1.5	0.9	2.235	0.72	0				
		White			1.65	0.41	0.12	0.22	1	1.65	0.41	0.12	0.22				
		Black		Plain	0.3	0.51	0.89	1.1	1	0.3	0.51	0.89	1.1				
		Red	MH	Glossy	0.6	1.49	0.48	0	3	1.8	4.47	1.44	0				
		Black		Glossy	0.71	0.39	0.6	0.3	5	3.55	1.95	3	1.5				
		Silver			1.15	0.59	1.27	0.24	5	5.75	2.95	6.35	1.2				
		Silver			1.15	0.59	1.27	0.24	2	2.3	1.18	2.54	0.48				
	RHS	Black		Rough	0.3	0.45	0.36	1.49	5	1.5	2.25	1.8	7.45				
		Red	MH	Glossy	0.6	1.49	0.48	0	2	1.2	2.98	0.96	0				
		Yellow	LL	Glossy	1.04	0.42	0.21	0.48	2	2.08	0.84	0.42	0.96				
		White			1.65	0.41	0.12	0.22	0.5	0.825	0.205	0.06	0.11				
		Silver			1.15	0.59	1.27	0.24	8	9.2	4.72	10.16	1.92				
		Black		Rough	0.3	0.45	0.36	1.49	5	1.5	2.25	1.8	7.45				
		Silver			1.15	0.59	1.27	0.24	2.5	2.875	1.475	3.175	0.6				
		White			1.65	0.41	0.12	0.22	2	3.3	0.82	0.24	0.44				
		Silver			1.15	0.59	1.27	0.24	1	1.15	0.59	1.27	0.24				
		Black		Plain	0.3	0.51	0.89	1.1	3	0.9	1.53	2.67	3.3				

(continued)

(continued)

Model	Side	Details			Texture Score (x)				n	xn				Total score = $\sum xn/\sum n$			
		Hue		Surf Texture	S	G	D	I		S	G	D	I	S	G	D	I
TVS Flame	LHS	Black		Glossy	0.71	0.39	0.6	0.3	1.5	1.065	0.585	0.9	0.45	0.55	0.75	0.57	1.17
		Red	MH	Glossy	0.6	1.49	0.48	0	2.5	1.5	3.725	1.2	0				
		Red	MH	Glossy	0.6	1.49	0.48	0	1.5	0.9	2.235	0.72	0				
		Red	MH	Glossy	0.6	1.49	0.48	0	1	0.6	1.49	0.48	0				
		White			1.65	0.41	0.12	0.22	0.5	0.825	0.205	0.06	0.11				
		Yellow	MH	Plain	1.01	0.77	0.86	0.24	1	1.01	0.77	0.86	0.24				
		Black		Rough	0.3	0.45	0.36	1.49	1	0.3	0.45	0.36	1.49				
		Black		Rough	0.3	0.45	0.36	1.49	7	2.1	3.15	2.52	10.43				
		White			1.65	0.41	0.12	0.22	2	3.3	0.82	0.24	0.44				
		Red	MH	Glossy	0.6	1.49	0.48	0	2.5	1.5	3.725	1.2	0				
		Black		Glossy	0.71	0.39	0.6	0.3	2	1.42	0.78	1.2	0.6				
		Black		Rough	0.3	0.45	0.36	1.49	8	2.4	3.6	2.88	11.92				
		Black		Rough	0.3	0.45	0.36	1.49	5	1.5	2.25	1.8	7.45				
	RHS	Black		Rough	0.3	0.45	0.36	1.49	7	2.1	3.15	2.52	10.43				
		Black		Glossy	0.71	0.39	0.6	0.3	4	2.84	1.56	2.4	1.2				
		Chrome		Glossy	2.02	2.8	3	2.78	4.5	9.09	12.6	13.5	12.51				
		Black		Rough	0.3	0.45	0.36	1.49	11	3.3	4.95	3.96	16.39				
		Red	MH	Glossy	0.6	1.49	0.48	0	3	1.8	4.47	1.44	0				
		Black		Rough	0.3	0.45	0.36	1.49	7	2.1	3.15	2.52	10.43				

(continued)

(continued)

Model	Side	Details			Texture Score (x)				n	xn				Total score = $\sum xn/\sum n$			
		Hue		Surf Texture	S	G	D	I		S	G	D	I	S	G	D	I
Discover	LHS	Black		Glossy	0.71	0.39	0.6	0.3	2	1.42	0.78	1.2	0.6	0.76	0.56	0.59	0.69
		Red	MH	Glossy	0.6	1.49	0.48	0	3	1.8	4.47	1.44	0				
		White			1.65	0.41	0.12	0.22	1	1.65	0.41	0.12	0.22				
		Silver			1.15	0.59	1.27	0.24	5	5.75	2.95	6.35	1.2				
		Black		Plain	0.3	0.51	0.89	1.1	3	0.9	1.53	2.67	3.3				
		White			1.65	0.41	0.12	0.22	0.5	0.825	0.205	0.06	0.11				
		White			1.65	0.41	0.12	0.22	0.5	0.825	0.205	0.06	0.11				
		Black		Glossy	0.71	0.39	0.6	0.3	7	4.97	2.73	4.2	2.1				
		Black		Rough	0.3	0.45	0.36	1.49	6	1.8	2.7	2.16	8.94				
		Silver			1.15	0.59	1.27	0.24	5	5.75	2.95	6.35	1.2				
		White			1.65	0.41	0.12	0.22	1	1.65	0.41	0.12	0.22				
	RHS	Black		Rough	0.3	0.45	0.36	1.49	5	1.5	2.25	1.8	7.45				
		Black		Rough	0.3	0.45	0.36	1.49	1	0.3	0.45	0.36	1.49				
		Red	MH	Glossy	0.6	1.49	0.48	0	2	1.2	2.98	0.96	0				
		White			1.65	0.41	0.12	0.22	1	1.65	0.41	0.12	0.22				
		White			1.65	0.41	0.12	0.22	1	1.65	0.41	0.12	0.22				
		White			1.65	0.41	0.12	0.22	3	4.95	1.23	0.36	0.66				
		Black		Plain	0.3	0.51	0.89	1.1	1	0.3	0.51	0.89	1.1				
		Black		Rough	0.3	0.45	0.36	1.49	5	1.5	2.25	1.8	7.45				

(continued)

(continued)

Model	Side	Details			Texture Score (x)				n	xn				Total score = $\sum xn/\sum n$			
		Hue		Surf Texture	S	G	D	I		S	G	D	I	S	G	D	I
Apache	LHS	Red		Glossy	0.6	1.49	0.48	0	5	3	7.45	2.4	0	0.83	0.71	0.68	0.48
		Red		Glossy	0.6	1.49	0.48	0	5	3	7.45	2.4	0				
		Black		Glossy	0.71	0.39	0.6	0.3	3	2.13	1.17	1.8	0.9				
		Silver			1.15	0.59	1.27	0.24	5	5.75	2.95	6.35	1.2				
		Silver			1.15	0.59	1.27	0.24	3	3.45	1.77	3.81	0.72				
		Red		Glossy	0.6	1.49	0.48	0	2	1.2	2.98	0.96	0				
		Black		Plain	0.3	0.51	0.89	1.1	2	0.6	1.02	1.78	2.2				
		Red	MH	Glossy	0.6	1.49	0.48	0	3	1.8	4.47	1.44	0				
		White			1.65	0.41	0.12	0.22	0.5	0.825	0.205	0.06	0.11				
		Red	MH	Glossy	0.6	1.49	0.48	0	3	1.8	4.47	1.44	0				
		Black		Glossy	0.71	0.39	0.6	0.3	2.5	1.775	0.975	1.5	0.75				
		Black		Rough	0.3	0.45	0.36	1.49	6	1.8	2.7	2.16	8.94				
		Silver			1.15	0.59	1.27	0.24	6	6.9	3.54	7.62	1.44				
		White			1.65	0.41	0.12	0.22	1	1.65	0.41	0.12	0.22				
	RHS	Black		Rough	0.3	0.45	0.36	1.49	4	1.2	1.8	1.44	5.96				
		Black		Glossy	0.71	0.39	0.6	0.3	4	2.84	1.56	2.4	1.2				
		Silver			1.15	0.59	1.27	0.24	5	5.75	2.95	6.35	1.2				
		Silver			1.15	0.59	1.27	0.24	2.5	2.875	1.475	3.175	0.6				
		White			1.65	0.41	0.12	0.22	8	13.2	3.28	0.96	1.76				
		Black		Rough	0.3	0.45	0.36	1.49	7	2.1	3.15	2.52	10.43				
		Silver			1.15	0.59	1.27	0.24	3	3.45	1.77	3.81	0.72				
		Black		Plain	0.3	0.51	0.89	1.1	1	0.3	0.51	0.89	1.1				

(continued)

(continued)

Model	Side	Details		Texture Score (x)				n	xn				Total score = $\sum xn/\sum n$			
		Hue	Surf Texture	S	G	D	I		S	G	D	I	S	G	D	I
Pulsar	LHS	Black	Glossy	0.71	0.39	0.6	0.3	8	5.68	3.12	4.8	2.4	0.80	0.47	0.73	0.61
		White		1.65	0.41	0.12	0.22	1	1.65	0.41	0.12	0.22				
		Silver		1.15	0.59	1.27	0.24	2.5	2.875	1.475	3.175	0.6				
		Silver		1.15	0.59	1.27	0.24	4	4.6	2.36	5.08	0.96				
		Black	Plain	0.3	0.51	0.89	1.1	0.2	0.06	0.102	0.178	0.22				
		Black	Plain	0.3	0.51	0.89	1.1	3	0.9	1.53	2.67	3.3				
		Black	Glossy	0.71	0.39	0.6	0.3	4	2.84	1.56	2.4	1.2				
		Black	Rough	0.3	0.45	0.36	1.49	6	1.8	2.7	2.16	8.94				
		Silver		1.15	0.59	1.27	0.24	3	3.45	1.77	3.81	0.72				
		Silver		1.15	0.59	1.27	0.24	1.5	1.725	0.885	1.905	0.36				
		Black	Rough	0.3	0.45	0.36	1.49	1	0.3	0.45	0.36	1.49				
	RHS	Black	Rough	0.3	0.45	0.36	1.49	6	1.8	2.7	2.16	8.94				
		Black	Glossy	0.71	0.39	0.6	0.3	11	7.81	4.29	6.6	3.3				
		Silver		1.15	0.59	1.27	0.24	0.5	0.575	0.295	0.635	0.12				
		Silver		1.15	0.59	1.27	0.24	8.5	9.775	5.015	10.795	2.04				
		White		1.65	0.41	0.12	0.22	5	8.25	2.05	0.6	1.1				
		Black	Rough	0.3	0.45	0.36	1.49	4.5	1.35	2.025	1.62	6.705				
		Silver		1.15	0.59	1.27	0.24	2.5	2.875	1.475	3.175	0.6				
		Black	Plain	0.3	0.51	0.89	1.1	1	0.3	0.51	0.89	1.1				

(continued)

(continued)

Model	Side	Details			Texture Score (x)				n	xn				Total score = ∑xn/∑n			
		Hue		Surf Texture	S	G	D	I		S	G	D	I	S	G	D	I
Splendor+	LHS	White			1.65	0.41	0.12	0.22	0.5	0.825	0.205	0.06	0.11	0.90	0.48	0.69	0.51
		Blue	MH	Glossy	0.92	0.98	0.71	0.12	0.5	0.46	0.49	0.355	0.06				
		White			1.65	0.41	0.12	0.22	3	4.95	1.23	0.36	0.66				
		Blue	MH	Glossy	0.92	0.98	0.71	0.12	0.5	0.46	0.49	0.355	0.06				
		Black		Glossy	0.71	0.39	0.6	0.3	4.5	3.195	1.755	2.7	1.35				
		Black		Glossy	0.71	0.39	0.6	0.3	3	2.13	1.17	1.8	0.9				
		Black		Glossy	0.71	0.39	0.6	0.3	2	1.42	0.78	1.2	0.6				
		Silver			1.15	0.59	1.27	0.24	9	10.35	5.31	11.43	2.16				
		Blue	MH	Glossy	0.92	0.98	0.71	0.12	0.5	0.46	0.49	0.355	0.06				
		Black		Glossy	0.71	0.39	0.6	0.3	13	9.23	5.07	7.8	3.9				
		Silver			1.15	0.59	1.27	0.24	2	2.3	1.18	2.54	0.48				
		Silver			1.15	0.59	1.27	0.24	2	2.3	1.18	2.54	0.48				
	RHS	White			1.65	0.41	0.12	0.22	1.5	2.475	0.615	0.18	0.33				
		White			1.65	0.41	0.12	0.22	0.5	0.825	0.205	0.06	0.11				
		Black		Rough	0.3	0.45	0.36	1.49	6	1.8	2.7	2.16	8.94				
		Black		Glossy	0.71	0.39	0.6	0.3	1	0.71	0.39	0.6	0.3				
		Black		Glossy	0.71	0.39	0.6	0.3	1	0.71	0.39	0.6	0.3				
		Silver			1.15	0.59	1.27	0.24	9.5	10.925	5.605	12.065	2.28				
		Black		Rough	0.3	0.45	0.36	1.49	9	2.7	4.05	3.24	13.41				
		White			1.65	0.41	0.12	0.22	7	11.55	2.87	0.84	1.54				
		Red	MH	Plain	0.45	0.48	1.1	0.89	0.5	0.225	0.24	0.55	0.445				
		Black		Plain	0.3	0.51	0.89	1.1	2	0.6	1.02	1.78	2.2				
		Silver			1.15	0.59	1.27	0.24	1.5	1.725	0.885	1.905	0.36				

(continued)

(continued)

Model	Side	Details			Texture Score (x)				n	xn				Total score = ∑xn/∑n			
		Hue		Surf Texture	S	G	D	I		S	G	D	I	S	G	D	I
Splendor NXG	LHS	White			1.65	0.41	0.12	0.22	1	1.65	0.41	0.12	0.22	0.76	0.57	0.79	0.72
		White			1.65	0.41	0.12	0.22	1	1.65	0.41	0.12	0.22				
		Chrome		Glossy	2.02	2.8	3	2.78	1	2.02	2.8	3	2.78				
		Red	MH	Glossy	0.6	1.49	0.48	0	2	1.2	2.98	0.96	0				
		White			1.65	0.41	0.12	0.22	0.5	0.825	0.205	0.06	0.11				
		Red	MH	Glossy	0.6	1.49	0.48	0	1	0.6	1.49	0.48	0				
		Black		Glossy	0.71	0.39	0.6	0.3	1.5	1.065	0.585	0.9	0.45				
		Black		Glossy	0.71	0.39	0.6	0.3	2.5	1.775	0.975	1.5	0.75				
		Black		Glossy	0.71	0.39	0.6	0.3	4	2.84	1.56	2.4	1.2				
		White			1.65	0.41	0.12	0.22	0.5	0.825	0.205	0.06	0.11				
		Red	MH	Plain	0.45	0.48	1.1	0.89	0.5	0.225	0.24	0.55	0.445				
		Black		Plain	0.3	0.51	0.89	1.1	6	1.8	3.06	5.34	6.6				
		Black		Plain	0.3	0.51	0.89	1.1	6	1.8	3.06	5.34	6.6				
		Silver			1.15	0.59	1.27	0.24	4.5	5.175	2.655	5.715	1.08				
	RHS	White			1.65	0.41	0.12	0.22	0.5	0.825	0.205	0.06	0.11				
		Red	LL	Glossy	1.16	0.21	0.36	0.27	0.5	0.58	0.105	0.18	0.135				
		Black		Rough	0.3	0.45	0.36	1.49	6	1.8	2.7	2.16	8.94				
		Black		Plain	0.3	0.51	0.89	1.1	4	1.2	2.04	3.56	4.4				
		Silver			1.15	0.59	1.27	0.24	8	9.2	4.72	10.16	1.92				
		Black		Rough	0.3	0.45	0.36	1.49	4	1.2	1.8	1.44	5.96				
		Silver			1.15	0.59	1.27	0.24	3	3.45	1.77	3.81	0.72				
		White			1.65	0.41	0.12	0.22	3.5	5.775	1.435	0.42	0.77				
		Red	MH	Plain	0.45	0.48	1.1	0.89	1	0.45	0.48	1.1	0.89				
		Black		Plain	0.3	0.51	0.89	1.1	1	0.3	0.51	0.89	1.1				

(continued)

(continued)

Model	Side	Details			Texture Score (x)				n	xn				Total score = $\sum xn/\sum n$			
		Hue		Surf Texture	S	G	D	I		S	G	D	I	S	G	D	I
Passion	LHS	Red	MH	Glossy	0.6	1.49	0.48	0	2	1.2	2.98	0.96	0	0.70	0.72	0.65	0.65
		Red	MH	Glossy	0.6	1.49	0.48	0	3	1.8	4.47	1.44	0				
		Black		Glossy	0.71	0.39	0.6	0.3	2	1.42	0.78	1.2	0.6				
		Black		Glossy	0.71	0.39	0.6	0.3	10	7.1	3.9	6	3				
		Red	MH	Glossy	0.6	1.49	0.48	0	0.5	0.3	0.745	0.24	0				
		Black		Glossy	0.71	0.39	0.6	0.3	4	2.84	1.56	2.4	1.2				
		Black		Rough	0.3	0.45	0.36	1.49	6	1.8	2.7	2.16	8.94				
		Black		Rough	0.3	0.45	0.36	1.49	4	1.2	1.8	1.44	5.96				
		Silver			1.15	0.59	1.27	0.24	1	1.15	0.59	1.27	0.24				
		Chrome		Glossy	2.02	2.8	3	2.78	1.5	3.03	4.2	4.5	4.17				
		White			1.65	0.41	0.12	0.22	0.5	0.825	0.205	0.06	0.11				
		Black		Rough	0.3	0.45	0.36	1.49	0.5	0.15	0.225	0.18	0.745				
	RHS	Black		Rough	0.3	0.45	0.36	1.49	5.5	1.65	2.475	1.98	8.195				
		Red		Glossy	0.6	1.49	0.48	0	2	1.2	2.98	0.96	0				
		Red	MH	Glossy	0.6	1.49	0.48	0	6	3.6	8.94	2.88	0				
		Silver			1.15	0.59	1.27	0.24	4	4.6	2.36	5.08	0.96				
		Silver			1.15	0.59	1.27	0.24	3	3.45	1.77	3.81	0.72				
		White			1.65	0.41	0.12	0.22	2.5	4.125	1.025	0.3	0.55				
		Black		Rough	0.3	0.45	0.36	1.49	4	1.2	1.8	1.44	5.96				
		Silver			1.15	0.59	1.27	0.24	2	2.3	1.18	2.54	0.48				
		Silver			1.15	0.59	1.27	0.24	1	1.15	0.59	1.27	0.24				
		Black		Plain	0.3	0.51	0.89	1.1	1	0.3	0.51	0.89	1.1				

(continued)

(continued)

Model	Side	Details			Texture Score (x)				n	xn				Total score = $\sum xn/\sum n$			
		Hue		Surf Texture	S	G	D	I		S	G	D	I	S	G	D	I
CD Dlx	LHS	Red		Glossy	0.6	1.49	0.48	0	0.5	0.3	0.745	0.24	0	0.70	1.33	0.72	0.76
		Red		Glossy	0.6	1.49	0.48	0	1	0.6	1.49	0.48	0				
		Red		Glossy	0.6	1.49	0.48	0	4	2.4	5.96	1.92	0				
		Red		Glossy	0.6	1.49	0.48	0	1	0.6	1.49	0.48	0				
		Black		Glossy	0.71	0.39	0.6	0.3	2	1.42	0.78	1.2	0.6				
		Black		Plain	0.3	0.51	0.89	1.1	3.5	1.05	1.785	3.115	3.85				
		Black		Plain	0.3	0.51	0.89	1.1	2	0.6	1.02	1.78	2.2				
		Black		Rough	0.3	0.45	0.36	1.49	9	2.7	4.05	3.24	13.41				
		Silver			1.15	0.59	1.27	0.24	6	6.9	3.54	7.62	1.44				
	RHS	Black		Rough	0.3	0.45	0.36	1.49	6	1.8	2.7	2.16	8.94				
		White			1.65	0.41	0.12	0.22	1	1.65	0.41	0.12	0.22				
		Silver			1.15	0.59	1.27	0.24	1.5	1.725	0.885	1.905	0.36				
		White			1.65	0.41	0.12	0.22	3	4.95	1.23	0.36	0.66				
		Black		Rough	0.3	0.45	0.36	1.49	6.5	1.95	2.925	2.34	9.685				
		Silver			1.15	0.59	1.27	0.24	9	10.35	5.31	11.43	2.16				
		Red		Plain	0.45	0.48	1.1	0.89	1	0.45	0.48	1.1	0.89				
		Silver			1.15	0.59	1.27	0.24	2.5	2.875	1.475	3.175	0.6				
		Black		Plain	0.3	0.51	0.89	1.1	1	0.3	0.51	0.89	1.1				

(continued)

(continued)

Model	Side	Details		Texture Score (x)				n	xn				Total score = $\sum xn/\sum n$			
		Hue	Surf Texture	S	G	D	I		S	G	D	I	S	G	D	I
CBZ	LHS	Black	Plain	0.3	0.51	0.89	1.1	4	1.2	2.04	3.56	4.4	0.75	0.51	0.89	0.70
		White		1.65	0.41	0.12	0.22	0.5	0.825	0.205	0.06	0.11				
		Red	Plain	0.45	0.48	1.1	0.89	1	0.45	0.48	1.1	0.89				
		Black	Glossy	0.71	0.39	0.6	0.3	5	3.55	1.95	3	1.5				
		Black	Plain	0.3	0.51	0.89	1.1	2.5	0.75	1.275	2.225	2.75				
		Silver		1.15	0.59	1.27	0.24	2	2.3	1.18	2.54	0.48				
		Silver		1.15	0.59	1.27	0.24	3	3.45	1.77	3.81	0.72				
		Black	Plain	0.3	0.51	0.89	1.1	4	1.2	2.04	3.56	4.4				
		White		1.65	0.41	0.12	0.22	1	1.65	0.41	0.12	0.22				
		Black	Glossy	0.71	0.39	0.6	0.3	4	2.84	1.56	2.4	1.2				
		Silver		1.15	0.59	1.27	0.24	5	5.75	2.95	6.35	1.2				
		Silver		1.15	0.59	1.27	0.24	5	5.75	2.95	6.35	1.2				
	RHS	Black	Rough	0.3	0.45	0.36	1.49	13	3.9	5.85	4.68	19.37				
		White		1.65	0.41	0.12	0.22	0.5	0.825	0.205	0.06	0.11				
		Silver		1.15	0.59	1.27	0.24	1	1.15	0.59	1.27	0.24				
		Silver		1.15	0.59	1.27	0.24	10	11.5	5.9	12.7	2.4				
		Silver		1.15	0.59	1.27	0.24	3.5	4.025	2.065	4.445	0.84				
		Red	Plain	0.45	0.48	1.1	0.89	1	0.45	0.48	1.1	0.89				
		Black	Rough	0.3	0.45	0.36	1.49	3.5	1.05	1.575	1.26	5.215				
		Silver		1.15	0.59	1.27	0.24	2	2.3	1.18	2.54	0.48				
		Red	Plain	0.45	0.48	1.1	0.89	2	0.9	0.96	2.2	1.78				
		Black	Plain	0.3	0.51	0.89	1.1	2	0.6	1.02	1.78	2.2				

(continued)

(continued)

Model	Side	Details			Texture Score (x)				n	xn				Total score = $\sum xn/\sum n$			
		Hue		Surf Texture	S	G	D	I		S	G	D	I	S	G	D	I
Glamor	LHS	Black		Glossy	0.71	0.39	0.6	0.3	3.5	2.485	1.365	2.1	1.05	0.67	0.40	0.71	0.64
		Black		Glossy	0.71	0.39	0.6	0.3	1	0.71	0.39	0.6	0.3				
		Black		Glossy	0.71	0.39	0.6	0.3	3	2.13	1.17	1.8	0.9				
		Red		Glossy	0.6	1.49	0.48	0	3	1.8	4.47	1.44	0				
		Red		Glossy	0.6	1.49	0.48	0	1	0.6	1.49	0.48	0				
		Silver			1.15	0.59	1.27	0.24	2	2.3	1.18	2.54	0.48				
		Black		Glossy	0.71	0.39	0.6	0.3	3	2.13	1.17	1.8	0.9				
		Black		Rough	0.3	0.45	0.36	1.49	6	1.8	2.7	2.16	8.94				
		Silver			1.15	0.59	1.27	0.24	5	5.75	2.95	6.35	1.2				
	RHS	Black		Rough	0.3	0.45	0.36	1.49	6	1.8	2.7	2.16	8.94				
		Red		Glossy	0.6	1.49	0.48	0	7	4.2	10.43	3.36	0				
		Silver			1.15	0.59	1.27	0.24	9	10.35	5.31	11.43	2.16				
		Black		Rough	0.3	0.45	0.36	1.49	5	1.5	2.25	1.8	7.45				
		Black		Plain	0.3	0.51	0.89	1.1	4	1.2	2.04	3.56	4.4				
		White			1.65	0.41	0.12	0.22	0.5	0.825	0.205	0.06	0.11				
		White			1.65	0.41	0.12	0.22	0.5	0.825	0.205	0.06	0.11				
		Black		Plain	0.3	0.51	0.89	1.1	2	0.6	1.02	1.78	2.2				

Appendix 4: Weighted Emotion Scores for Texture

Model	S	G	D	I	S	G	D	I
Hunk	0.43	1.08	0.89	0.95	0.98	2.43	2.00	2.15
TVS Star	0.87	0.64	0.78	0.45	1.96	1.45	1.76	1.02
TVS Flame	0.55	0.75	0.57	1.17	1.24	1.68	1.28	2.63
Discover	0.76	0.56	0.59	0.69	1.72	1.27	1.33	1.56
Apache	0.83	0.71	0.68	0.48	1.87	1.61	1.53	1.09
Pulsar	0.80	0.47	0.73	0.61	1.81	1.07	1.64	1.37
Splendor+	0.90	0.48	0.69	0.51	2.04	1.08	1.56	1.16
Splendor NXG	0.76	0.57	0.79	0.72	1.71	1.29	1.79	1.62
Passion	0.70	0.72	0.65	0.65	1.59	1.63	1.47	1.48
CD Dlx	0.70	1.33	0.72	0.76	1.59	3.00	1.62	1.72
CBZ	0.75	0.51	0.89	0.70	1.69	1.15	2.00	1.57
Glamor	0.67	0.40	0.71	0.64	1.50	0.90	1.59	1.44

Chapter 14
Design and the Role of Cultures

Culture is the defining entity of current times. In fact, it has always been the defining entity of contemporary times. Come to think of it, the history of humanity has been the history of migration and migration of people across societies and continents leads to the amalgamation of cultures which has been the moving force for design throughout history. Whether it is the evolution of Gothic art and architecture or the evolution of Arabic architecture in India, every milestone in the history of design can be attributed to the fusion of cultures resulting from migration and the most visible manifestation of this can be seen on design of artifacts—pottery, sculpture, fine arts, architecture, textiles and of course, automobiles. When the Japanese manufacturer Honda designed Splendor and Passion motorcycles for India, the Indian lyricism was consciously built into the Japanese design. Similarly, when, Harley crossed the American shores for the first time to establish its factory in India, the design thinking also started adjusting itself to Indian sensibility (Fig. 14.1).

Now, culture has its own dialectics which generates two opposing tendencies in reaction to the migration of cultures. One tendency, generally known as "liberalism" supports easy fusion of cultures while the other one generally called "conservatism" opposes this fusion and tends to protect the indigenous cultures. These two tendencies generate their own narratives and the clash of these narratives is dictating the politics, sociology, economics, literature, arts, and industry almost everywhere in the world. Needless to say, the world of design cannot escape this clash of narratives especially because it is affected by all of these. The conservative narrative reflects in the form of isolationist designs whereas the liberal narrative is found in generous doses of fusion. The interesting facet of this clash of narrative is the peaceful coexistence of both these narratives in the field of design whereas in social and political arena it finds expression in violent struggles. When I say peaceful coexistence of narratives, what I mean is that you will find both these narratives sometimes by same designers, sometimes same manufacturers and most surprisingly sometimes in same products.

S. Chandra, *Aesthetics: Quantification and Deconstruction*,
Design Science and Innovation, https://doi.org/10.1007/978-981-10-6235-3_14

Fig. 14.1 Harley Davidson Street (India) on (*left*) and Roadster (US) on (*right*). Note the slow curvature of fuel tank, high unity between visor, fuel tank, seat and rear cowl and soft contours of seat in Street against flat sided fuel tank, the absence of unity and sharper contours of seat in Roadster. Both these vehicle have almost the same engineering architecture but have been visually altered to suit cultural differences

But, before we discuss further the characteristics of culture and its relationship with design, let us first understand what exactly this widely used term means. I, in fact, have a very unflattering view of culture. In my view, it is the invisible superstructure of society representing its most tyrannical power dynamics and expressing itself through various avenues of expression like fine arts, literature, and design. Let us view the famous painting by Goya, "*The 3rd of May*" (Fig. 14.2). Though this work is one of the finest expressions of the democratic churning in Europe, a critical examination reveals the deep social fault-lines. We find the innocent democratic resistance in pure white whereas the brutal suppressive regime is shown in black color. The same social fault-lines get translated into the language

Fig. 14.2 *The 3rd of May* by Goya

of aesthetics where white becomes the color of serenity and purity and black starts representing intimidation and domination.

14.1 Migration

Migration of design influences due to cultural migration has been a historical phenomenon as we have discussed already. To analyze it further, let us consider building architecture. Indian building architecture has been affected by this migration many times over. The Gothic architecture of Europe evolved itself in form of Arabic architecture and during the great architectural renaissance in India during the Moghul era, the Indian architecture got a dash of Gothic influence in form of double arches used in the design of entrances, windows, and domes. This style later influenced the design of Hindu temples as seen in the interiors of Govind Dev temple in Mathura and Chaturbhuj temple in Orchha (Fig. 14.3).

Another example of architectural migration is the influence of British architecture in India. It started with the colonial architecture used by the British and slowly crept into Indian style of structures even to the extent of temples. The Victoria Terminus railway station was built by the British in Victorian Gothic style and we see the impact of this style on the domes of Akshardham temple built in late twentieth century. Similarly, you can see the influence of Arabic architecture in Europe in Palatine Chapel in Sicily inspired from the art of Fatimids (Fig. 14.4).

Coming back to motorcycles, the initial phase of motorcycles in India (1961–1984) was influenced by the classical architecture. With the onset of Japanese invasion in 1984, the minimalist school of Japan slowly took over, which again under the influence of Indian lyricism converged into a hybrid version of minimalism. We will discuss in detail the effects of Indian lyricism on design. Meanwhile, the latest flush of European and American bike companies setting up their shops in India has started leaving its imprint in both directions. We have already seen this imprint on Harley in earlier pages of this chapter.

Fig. 14.3 Moghul era architecture (*left*), Govind Dev temple in Mathura (*Center*) and Chaturbhuj temple in Orchha (*Right*)

Fig. 14.4 Victoria Terminus Railway Station, Mumbai (*left*), Akshardham Temple Delhi (*Center*) and Palatine Chapel in Sicily, Italy (*Right*)

14.2 Gender

Now the most eloquent example of this tyrannical power dynamics is the gender equation prevalent in almost every part of the world and in all types of societies. Feminism is the counter-discourse of this dynamics. Without going into history and etymology of feminism, it would be relevant to posit together the Western and Indian concepts of feminism. A question can be asked as to why at all the two concepts should be different as long as feminism is about fighting for equality of women. Though at the outset, throughout the world, feminism is about fighting inequalities in a male-centric world, the contexts of this fight have been changing in different societies. The theoretical foundations of feminism as we know today, can be found in the works of Simone De Beauvoir and Germaine Greer, where Beauvoir's "The Second Sex" traces the seeds of gender oppression in economic social structures, Greer's "The Female Eunuch" questions the value systems and prejudices as formulated from the standpoint of men. This, in my view, is a very important basis to define feminism in simple terms—women's refusal to be defined by men.

Again, what is stated is the expression of what often is not stated, i.e., the sub-text of our belief systems because prejudices often lurk in sub-texts. These sub-texts are expressed through various avenues and design of artifacts like two-wheelers which touch our lives every day. If we look carefully at the designs of motorcycles and scooters through the years, they tell a fascinating tale of the changing (or not changing to be precise) social narratives (Figs. 14.5 and 14.6).

With a naked construction, higher cc high powered engine and fuel tanks with overt convexity, motorcycles in pre-liberalization India represented the brute dominance of feudal patriarchy (Fig. 14.7). On the other end, scooters having 150 cc engines with relatively lower torque and power and relatively higher comfort and convenience represented the needs of self-proclaimed urban sophistication, which exposes itself when you notice in this whole scheme of things, the peripheral role of women restricted to the pillion seat.

We can notice the transformation in the character of motorcycles from pre-80 period to post-90. From a high cc, high power brute, it changed to 150 cc lower powered avatar with fuel-efficiency as the theme song suitable to the urban commuter. But the styling is distinctly male with either predominantly convex surfaces

Fig. 14.5 Motorcycles (Pre-80)

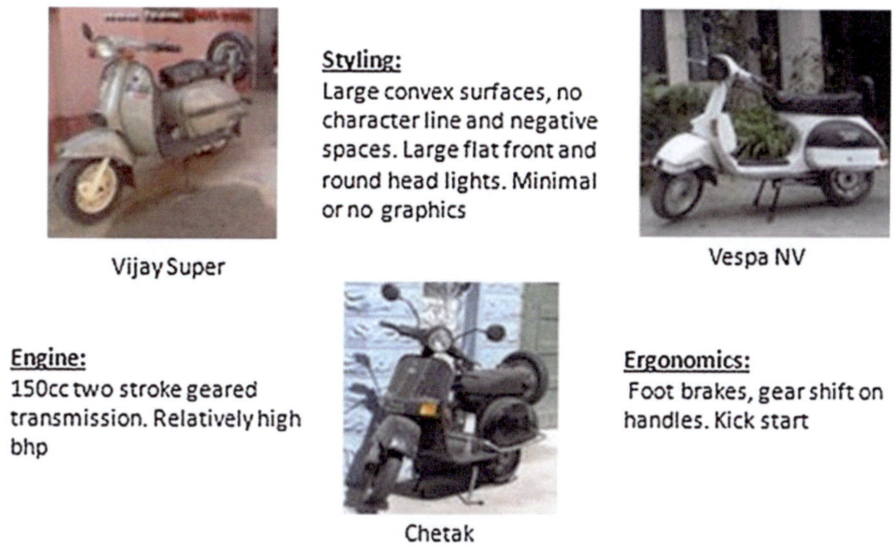

Fig. 14.6 Scooters (Pre-80)

or overt positive–negative combinations with sharp character lines. In short, it retained the masculine character through its triumphant journey to urban landscape (refer Figs. 14.8, 14.9 and 14.10).

Here we can clearly see the contours of design and sociocultural evolution getting into each other's territory. Pushed from its urban domination and forced to absolute extinction, the Indian scooter made a remarkable comeback in form of "Honda-Activa" with a softer convex or positive–negative combinations of surfaces and variometric transmissions with self-start appealing to large chunk of urban

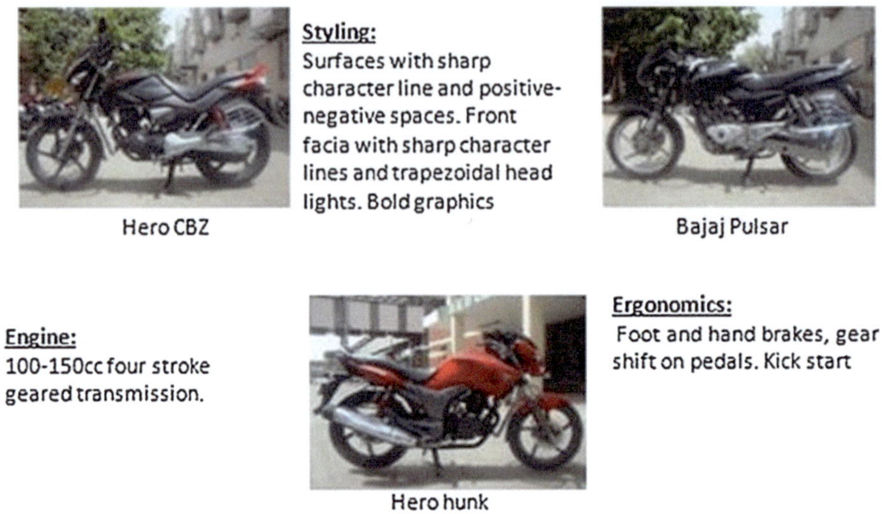

Fig. 14.7 Motorcycles (Post-90)

Fig. 14.8 Scooters-Post-2000 (Female)

women. A keen observer of Indian two-wheeler market will tell you that Activa was initially not aimed at women in particular but it was positioned as an urban vehicle of convenience. But the feminine identification acquired a life of its own and it got further accentuated with Hero-Pleasure which explicitly positioned itself as "JUST FOR HER". This is where the male–female binary got sharp and clear with

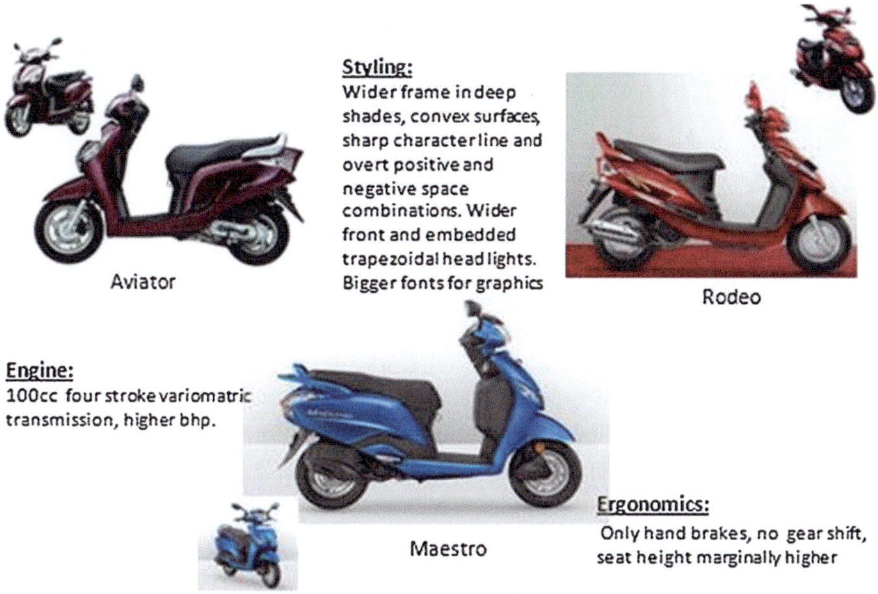

Fig. 14.9 Scooters-Post-2000 (Male)

Fig. 14.10 Scooters-Post-2000 (Unisex)

motorcycle and scooter representing the two. Not only this, with passage of time the grammar of scooter design got nuanced and a fine distinction between male, female and unisex categories emerged, which we will discuss in detail. Before we analyze the grammar, we look at the comparison of design details to study the genderization of motorcycle and scooter design.

The table (Table 14.1) compares the styling parameters like color, form, graphics, lights and aesthetic parameters like unity and dynamism considering the chronological perspective.

Hidden in technical text of the table is the social narrative of India—the sub-text of male domination. In pre-liberalization India before the 90s, motorcycle and scooter represented two clear points in the social map—both represented the male-centric society. In this age of socialist pretensions, scooter was a symbol of middle-class aspirations of urban India. But despite these socialist and utilitarian pretensions, there was an unmistakable imprint of dictating masculine paradigm. At the same time, the rural and subaltern India was largely feudal and the language of feudal dominance was expressed through the styling of motorcycles and fuel tanks with overt convexity. So both motorcycles and scooters were two faces of the same male-chauvinistic order and women were essentially restricted to the periphery.

By the time scooter reappeared on the scene, the motorcycle had completely captured the urban landscape and forced scooter to make an unceremonious exit. By this time, urban population had a large population of working women and the Indian social narrative did not allow the motorcycle to be the usual mode of transport for them.

Engine Power: Observing Table 14.2 again perpetuates the same narrative. In the pre-90 scenario, scooter and motorcycle were two sides of the same male-centric world. Scooter, of course, surfaced as an alternative for a modern woman but the dynamics remained unchanged. For a woman, power was secondary. Convenience and a hassle free ride were what really, mattered for them.

Ergonomics: This is where the inherent biases of patriarchy are thoroughly exposed as we can observe in Table 14.3. The unisex scooter is meant for a family, where all members of a family irrespective of gender ride it. But the styling, the power, the ergonomics—everything is designed to cater to the male user—the ultimate decision-maker. What actually separates the unisex from male is the convenience added to it and the overt aggression in styling is slightly toned down.

Grammar: Design is the text of the sociological narrative that motorcycles and scooters generate. This text is the outcome of the social and political sub-text and the tool to translate the sub-text into text is the grammar of aesthetics, which itself is an outcome of the civilizational evolution of a society. The grammar of design that emerges out of the analysis of design data clearly sketches the parameters of gender equations in Indian society. The emotional parameters of design get translated into engineering parameters as shown in the following table (Table 14.4).

Table 14.1 Comparison of styling

Parameter		Pre-80 motorcycle	Post-90 motorcycle	Scooter			
				Pre-90	Post-90 female	Post-90 male	Post-90 unisex
Color		Low-value high chroma, chrome	Low-value high chroma, chrome	High-value low chroma, no chrome	High-value low chroma, no chrome	Low-value high chroma, no chrome	←
Form	Side	Convex surfaces	High convex or ± combination	Large convex, no chamfers	Slim, mild ± combination	Overt ± combination, sharp chamfer	←
	Front	No visor	Visor/windshield with convexity	Large flat, no chamfer	Pointed inclined surface, soft chamfer	Pointed inclined surface, sharp chamfer	←
Graphics		No graphics	Overt graphics or bold emblems	No graphics	Nonintrusive, small cursive fonts		←
Lights		Round	Trapezoidal	Round	Cute trapezoidals	Bold trapezoidals	←
Unity and Dynamism		Low unity, low dynamism	High unity, high dynamism	Low unity, low dynamism	Higher unity and dynamism	←	←

Table 14.2 Comparison of Engine Power

Parameter	Pre-80 motorcycle	Post-90 motorcycle	Scooter			
			Pre-90	Post-90 female	Post-90 male	Post-90 unisex
Engine capacity	350–500 cc	100–200 cc	150	100	100–110	100–111
Power at Crankshaft (BHP)	Above 18 bhp	7.5–15 bhp	7.48–9.84	5–7	8–8.6	8–8.2

Table 14.3 Comparison of ergonomics

Parameter		Scooter		
		Post-90 female	Post-90 male	Post-90 unisex
Seat height (mm)		760–762	769–790	765
Anthropological crotch height	Male	765		
	Female	733		

An analysis of male, female and unisex scooters engineering stylistic features and their emotional and social interpretation, which, philosophically can be termed as sub-text of design, can be expressed as the following diagram (Fig. 14.11).

As we reveal the unspoken prejudices lurking in the inner layers of design, we see the true face of a society, which on the surface looks egalitarian and equitable providing equal opportunities across the gender-divide. But just scratch the surface and we find the age-old prejudices and social stereotypes still reigning supreme. We, find supremacy and aggression scoring highest for male scooters and lowest for female scooters. Not only this, the supposedly feminine virtues of vitality and fun-factor scoring highest for female scooters, thus further accentuating the stereotypes.

Table 14.4 Grammar of Aesthetics versus Engineering Parameters

Supremacy	Recognition	Aggression	Vitality	Fun factor	Functionality
Visual contrast	Exclusivity of features	Dynamism of surfaces	Intricate detailing	High-value high saturation colors	Useful features
Robustness of frame	Cosmetic detailing	Sharp polarization between positive and negative spaces	Sharp fillets between elements	Uncluttered yet dynamic surfaces	Comfortable ergonomics
Sharp polarization between positive and negative spaces			Bold colors		High durability

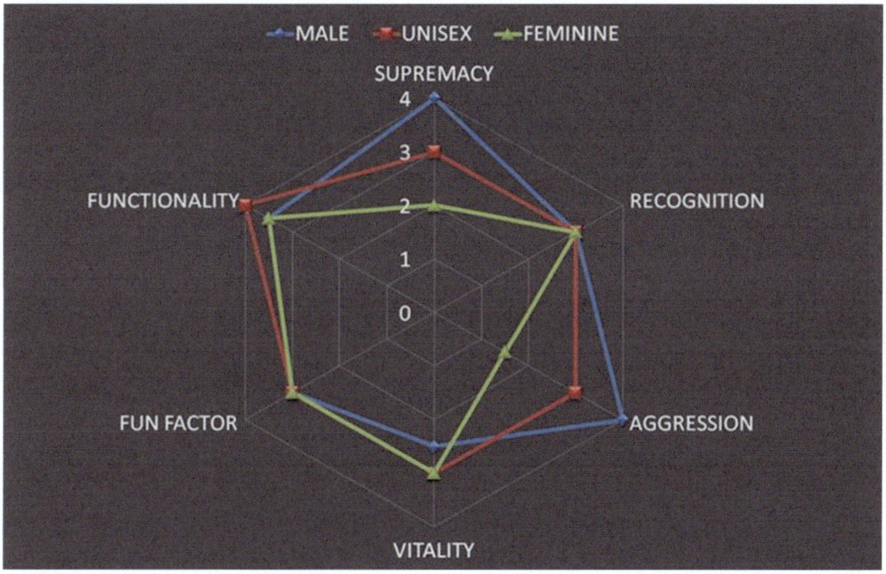

Fig. 14.11 Aesthetics versus gender

But, the real devil is hiding in unisex scooters, which reveal the harsh realities of Indian families. As unisex scooters are supposed to be functional and to be used by all members of family irrespective of gender, particularly where the female member is earning as well, when it comes to decision-making, it is the man, who is the ultimate arbiter. So, the scooters have to reflect his emotional and ergonomic needs.

Reflecting the societies with multi-layered hierarchies in-built in the power equation of gender, the two-wheelers also have spun a complex web of sub-texts. Compared to Europe, where scooters do not have feminine connotation and US where motorcycles do not reflect the economic middle ground, India represents a distinct social equation. In the overall social ladder, motorcycles and scooters represent the middle ground—above bicycles and mopeds representing the underclass and below the car which represents the elite. This middle ground has a multi-layered gender-divide, where motorcycle is distinctly male and scooter is for women. So motorcycle–scooter are the signifiers of the binary and male-female the signified. Our movies, which actually represent the social realities in form of fantasies, show the larger-than-life macho heroes on motorcycles protecting the vulnerable heroine generally seen on scooters. Now, among the scooters, there is a fine distinction as discussed which reflect our prejudices and stereotypes. The larger point is, as long as the prejudices are there, the consequences of the prejudices will also be there—culminating in the male-centric belief systems.

14.3 Lyricism

The relationship between the society, culture, and design is expressed through lyricism inherent in a society and this is particularly true for India, where even the books on medical science (Ayurveda), law (Pramana Vartika) and astronomy (Leelavati by Bhaskaracharya) have been written in verse. To understand this, let us have a look on Indian society which has always been divided along three types of hierarchies-Administrative, caste, and feudal hierarchies running along identical lines, i.e., the people at top of the three hierarchies were the same. Similarly, People at the bottom of the three pyramids are also the same. Initially, cars were meant only for the select few among the urban rich. Motorcycles were meant for the rural rich and those forming the creamy layer of the urban middle class. This resulted in the fundamental social codes of rural dominance and urban quest for grace and class for motorcycle owners. In this period, motorcycles mostly acted as the instrument of romance in Indian movies like Bobby (1974), Andaaz (1970), and Ek Duje Ke Liye (1979). But the dominant social code for romance was not very explicit. In public life, one could see the wives and girlfriends on the pillion seat of the motorcycle in side facing posture. Overt public display of affection did not have wide social acceptability. Table 14.5 illustrates how the social codes of that period got translated into motorcycle design features.

Two things need to be underlined here. Since the volumes were miniscule and technology was primitive, the fuel tank contours and the seat geometry were designed to suit the highest common denominator and manufacturing feasibility. These factors restricted the scope for lyricism (here we define lyricism as a function of emotion score and flow, which are essentially the two elements of lyrics) and despite the differences in the fundamental social codes, and a deep-rooted quest for lyricism, the design remained low on lyricism. This leads to some more analysis of Indian social cauldron. Indian society has been a mixing pot of many cultural influences. Indian social life, before the influx of Islam and the Victorian values brought by the British, was very rich in romance and poetry. The classics of the first millennium by Kalidasa (Kumar Sambhavam, Abhijnan Shakuntalam), Jaidev (Geet Govindam), etc., sufficiently prove this besides the sculptures of Khajuraho, Konark, and Frescos of Ajanta. Not only this, in contrast to Islam and Christianity all Hindu Gods and Goddesses are supposed to live a healthy conjugal life. This observation brings a very sharp contradiction about pre-Islamic India. Whereas Indian society was deeply divided along caste and feudal lines, within the feudalistic compartments man–woman relationships were very open. But the Islamic and European influences left a deep scar on this openness. On one hand, India turned into a conservative society in terms of gender discourse, on the other hand feudal and caste lines not only remained intact but got deepened with hierarchical structure.

The ideals of French and Soviet revolution left a deep influence on the libertarian values driving the Indian freedom struggle. In 1961, when motorcycle manufacturing started in India, a fierce conflict was on between Victorian conservatism and

Table 14.5 Set of fundamental social codes vis-a-vis design of motorcycles

Social Category	Dominant social code	Motorcycle brand	Social interaction between user and motorcycle	Governing Rasa	Design feature
Rural affluent	Domination	Bullet	Show masculinity, brute power	Vira	Minimalist Design, Black and chrome colors, flat and muscular contours
Rural youth	Upward social mobility	Hero Honda CD100	Show stability and growth	Shaant	Simple, clean surfaces, robust styling, slightly stylish seat
Rural middle-class commuter	Conservative values	Hero Honda CD100, Street; Suzuki Max 100	Carrying family members and objects of trade	Shaant	Flat wide seat, matching muffler direction, fuel tank flat contours
Urban affluent youth	Individuality, masculinity	Yamaha RD 350	Convey a style statement and personality through the bike	Adhbhut	Powerful engine, flat seat, dual mufflers
Urban middle-class youth	Openness in relationships	Bajaj KB 125, Hero Honda Sleek, Suzuki Samurai, Shogun, Yamaha RX 135	Look good, show individuality and dynamism	Shringar	Simple and elegant surfaces, Seat with a slightly stylish contour
Urban middle-class commuter	Pride in family values, grace, and dignity	Hero Honda CD100, Street; Suzuki Max 100	Carrying wife and children, pride of possession	Shaant	Flat wide seat, Matching muffler direction, fuel tank flat contours

western liberalism. This conflict gets illustrated in a very interesting manner by two most enduring images of motorcycles in Indian movies. Unlike western biking movies like the "The Wild One", "The Great Escape", "The Leather Boys", and "The Hell Angels", where motorcycle becomes an extension of the protagonist's anger and aggression, Indian movies of that age have used motorcycles as vehicles of romance. In 1970 movie "Andaaz", the boy is driving Bullet 350 with the girl in pillion seat singing a chartbuster. Interestingly the girl is sitting in sidesaddle posture. This inhibited expression of love finds expression in flat contours of seat and fuel tank and horizontal muffler. In the path-breaking movie Bobby (1974), the girl, who claims to belong to twenty-first century, sits in parted legs position facing the front and nibbling the earlobes of the boy. This image was a precursor of future social behavior when this sitting posture became common in metros.

During 1984 to 2000, a much deeper development was taking place at sociopolitical level. The feudal, caste and administrative hierarchies, instead of remaining parallel as ever, started crisscrossing each other as the aspirations of the people on the fringes started coming to the fore. Inter-gender discourse became more forthright and upfront, at least in urban areas. The way a boy and a girl interacted with the motorcycle, was the extension of the Bobby phenomenon, with the difference that now, the motorcycle was a willing participant in the process. Take seat, fuel tank and body covers dynamic and unified profile of Hero Honda Sleek and the matching muffler inclination. All this social and moral upheaval produced another set of fundamental social codes as illustrated in Table 14.5.

There are two interesting observations from this period. First is related to Indian poetry. This period saw the formal end of lyrical poetry in Indian literature. As a natural corollary, poetry got completely detached from mass discourse. Consequently, music, specifically film music remained Indian peoples only diet available for their deep-rooted hunger for lyricism. Another aspect of this development was technological. By this time, new kinds of paintable plastics like ABS had evolved allowing freedom of design for body parts. Another technological evolution was CAD/CAM, liberating the designers from constraints of complexities. Fuel tank shapes with generous combination of positive and negative spaces were otherwise impossible. Advances in lighting technology like multi-reflectors further eased the life of designers, as lens shape was not a constraint. All these factors played a key role in improving the lyricism factor.

14.4 Accounting for Culture

With all these influences, how do we account for them, while deconstructing the emotions for a design? End of the day, to quote Pythagoras, everything boils down to numbers and we finally encounter the question—how do we account for culture in SGDI scores. The answer is simple, SGDI scores are not written in stone and they can change due to the influences that we have discussed. Let us revisit the fundamental relationship:

$$[r]_{s,g,d,i} = \left[\sum w_x r_x\right]_{s,g,d,i} \tag{3.7}$$

where

w_x Weightage for various design factors and sub-factors and
r_x emotion score due to factors and sub-factors.

Since the cultural sensibilities alter both these parameters, these two parameters need to be re-established in each new scenario. This means the whole process of surveys and compilation of emotion score has to be re-done. Sure, this is complicated. But culture, after all, is not something which can be denoted by a coefficient which can be readily multiplied as a factor.

MIX
Papier aus verantwortungsvollen Quellen
Paper from responsible sources
FSC® C105338

If you have any concerns about our products,
you can contact us on
ProductSafety@springernature.com

In case Publisher is established outside the EU,
the EU authorized representative is:
Springer Nature Customer Service Center GmbH
Europaplatz 3, 69115 Heidelberg, Germany

Printed by Libri Plureos GmbH
in Hamburg, Germany